The Dodgers Move West

THE

MOVE WEST

Neil J. Sullivan

New York   Oxford
OXFORD UNIVERSITY PRESS
1987

Oxford University Press

Oxford   New York   Toronto
Delhi   Bombay   Calcutta   Madras   Karachi
Petaling Jaya   Singapore   Hong Kong   Tokyo
Nairobi   Dar es Salaam   Cape Town
Melbourne   Auckland

and associated companies in
Beirut   Berlin   Ibadan   Nicosia

Copyright © 1987 by Neil J. Sullivan

Published by Oxford University Press, Inc.,
200 Madison Avenue, New York, New York 10016

Oxford is a registered trademark of Oxford University Press

All photographs courtesy of the Los Angeles Dodgers

Library of Congress Cataloging-in-Publication Data

Sullivan, Neil J., 1948–
The Dodgers move west.
Includes index.
1. Los Angeles Dodgers (Baseball team)—History.
2. Brooklyn Dodgers (Baseball team)—History.   3. Sports
and state—California.   4. Baseball—California—Manage-
ment.   5. Baseball—New York (N.Y.)—Management.
I. Title.
GV875.L6S84   1987   798.357'64'0979494   86-28633
ISBN 0-19-504366-9 (alk. paper)

Grateful acknowledgment is given for permission to quote from the following works:

Specified excerpts from *The Boys of Summer* by Roger Kahn, © 1971, 1972
by Roger Kahn. Reprinted by permission of Harper and Row, Publishers
Inc.

Excerpts from *Bums* by Peter Golenbock, © 1984 by Peter Golenbock. Reprinted by permission of the Putnam Publishing Group.

Excerpts from *The Power Broker* by Robert Caro, © 1974 by Robert Caro.
Originally published by Alfred A. Knopf, Inc. Reprinted by permission of
Random House, Inc.

2 4 6 8 9 7 5 3 1

Printed in the United States of America
on acid-free paper

To My Parents

Preface

The decision to move the Brooklyn Dodgers to Los Angeles after the 1957 season continues to be perhaps the most controversial franchise shift in sports history. The transfer was a kind of benchmark for modern sports: it heralded unprecedented national growth in the business of spectator sports, and it also triggered the kind of emotional reaction that in recent years has led cities and teams to battle in the various arenas of government over their respective rights and obligations.

In spite of all this subsequent activity, however, the Brooklyn Dodgers retain a romantic vitality that distinguishes them from other teams that have relocated. Unlike the Boston Braves, the Philadelphia Athletics, and the St. Louis Browns, the Dodgers were a very successful team, winning six pennants in their final eleven years in Brooklyn. They also enjoyed a close emotional tie with that borough, dating back to the time when Brooklyn was an independent city. In the 1950s the Dodgers were financially more prosperous than any other team in the major leagues, including the New York Yankees. This support and prosperity notwithstanding, the Dodgers moved just two years after capturing the World Series title that had persistently eluded them.

Several recent books have considered the history of the Brooklyn Dodgers and the impact their move had on residents of that community. They point to the widespread belief that the Dodgers left in anticipation of the hard times New York, and in particular Brooklyn, would suffer in the coming years. The feeling of abandonment continues to rankle Brooklynites.

Passions of this kind tend to cohere around a villain, and discussions of the Dodgers' move have focused on their president, Walter Francis O'Malley, the man who decided to move the team. A self-described

Tory, with jowls and a gravel voice, O'Malley is almost universally perceived as a Machiavelli who made no decision without a ruthlessly dispassionate analysis of how it would affect his profits. O'Malley as villain may offer some emotional satisfaction, but it is poor history. The principal flaw of this view is that it assumes O'Malley manipulated public officials in New York and in Los Angeles until he had wrung the most compelling offer from each city, only then determining whether to stay or to go. Thus O'Malley is thought to have shrewdly anticipated the profits the club has subsequently enjoyed in Los Angeles. It is further assumed that he suckered officials in Southern California into showering him with wealth at the expense of helpless taxpayers, and that he abandoned Brooklyn heartlessly after city officials in New York declined to give him a new stadium.

In proposing a different explanation, this book rests on the premise that Walter O'Malley was but one actor in two distinct political games played in New York and Los Angeles, and that the actions of both cities were greatly affected by turn-of-the-century decisions about the organization and execution of public power.

In 1897 Brooklyn lost its status as an independent city and became a borough of greater New York City. When the Board of Estimate was established in 1903 as the most important unit of city government, equal political representation was conferred on each borough, regardless of the size of its population. Brooklyn's interests could thus be frustrated by smaller rival communities with which it shared power. In the 1950s, when the Dodgers sought the help of city officials to remain in Brooklyn, cooperation was needed from other boroughs; none was forthcoming. Political control was then needed on the Board of Estimate; that was not possible.

In Los Angeles, earlier political decisions which eventually affected the Dodgers had less to do with the specific organization of city government than with ideologically conceived limitations on the exercise of discretion permitted to public officials. After bitter experiences with corruption in the nineteenth century, state reforms during the Progressive era placed public officials under unprecedented democratic control. One reform in particular, the referendum, threatened to nullify the Dodgers' move even after all the obstacles in the negotiation and ratification of a contract for land at Chavez Ravine apparently had been overcome. These earlier decisions helped shape the options the Dodgers faced in the 1950s. Although O'Malley's desire to replace Ebbets Field with a modern stadium precipitated reactions by government officials in New York and Los Angeles, he was as much the victim of political forces as their instigator.

This book also considers specific issues alleged about the Dodgers' move. Were there any grounds for expecting the tremendous support the Dodgers subsequently have enjoyed in Los Angeles? The city had tried and failed several times before to attract a major league franchise, and although several clubs thought the region held promise, it apparently was not considered a certain path to riches. The Athletics, Braves, and Browns—all of whom moved before the Dodgers—judged Milwaukee, Kansas City, and Baltimore to be more promising. O'Malley's move was a bold, even reckless gamble which no one before had been willing to take. For although the history of minor league baseball in Southern California was long on tradition with several teams enjoying loyal followings, at the end of the 1950s Pacific Coast League parks were hardly filled to capacity. Such interest in baseball as still existed might conceivably have been tied to local rivalries that wouldn't sustain a major league team, especially one from so remote a place as Brooklyn.

A second point made frequently about the Dodgers' move is that citizens and taxpayers in Los Angeles were exploited by the ill-advised decisions of their elected officials. This contention ignores the years of political and judicial challenges to the Dodgers' contract with the city. These issues are reviewed here as are the initial difficulties that hampered the negotiation and ratification of the Chavez Ravine contract in Los Angeles. Whatever spell O'Malley may have tried to cast over the officials of Southern California, the voters and taxpayers determined for themselves, through a referendum in 1958, whether the decision made on their behalf was the proper one. In addition to that democratic safeguard, individual taxpayers filed suit in local court to invalidate the agreement between the Dodgers and Los Angeles. In the midst of the dismal 1958 season, with the team ensconced in the National League cellar, a judge nullified the Chavez Ravine contract and set back the Dodgers' chances of ultimately settling in Los Angeles.

Finally, despite the belief of Brooklynites that they were abandoned after their years of loyal support, some evidence suggests that O'Malley left with grave reservations. Even assuming O'Malley to have been a shrewd calculator of self-interest, why did he exchange a community that had put so much wealth in his pocket for an untested market fifteen hundred miles beyond the previous western outpost of major league baseball?

A dispassionate analysis of the Dodgers' move to Los Angeles reveals Walter O'Malley, the putative villain, to have been considerably less circumspect than, as a business executive, he should have been. Indeed, the Dodgers' prosperity in California may ultimately rest on the simple fact that without adequate preparation Walter O'Malley as-

sumed a risk that other club owners had passed up. Had O'Malley
been aware of the obstacles Los Angeles would put in his path, he
might have stayed in New York or moved to another city. At some
critical moments the Dodgers' president was more lucky than cunning,
and the team's success in Los Angeles reflects his good fortune as well
as his business acumen.

Most importantly, the Dodgers' case demonstrates the significance
of a proper relationship between a community and its sports teams, a
relationship badly confused in recent years as private firms have made
claims on the public purse to finance stadiums. Cities have competed
with one another in providing arenas to attract or hold sports franchises, and many of these communities have expended resources in a
way that has made them hostages to club owners. The Dodgers and
Los Angeles have thrived as a model association of community and
ballclub. Citizens and the franchise both recognize that ball clubs are
team. Citizens and the franchise both recognize that ball clubs are businesses, not wards of the state, and that they should be responsible for
providing their own ballparks with limited government assistance. Curiously, in spite of the Dodgers' success in Los Angeles, no baseball
team has since tried to build a stadium through private financing. The
model association has been widely ignored.

Acknowledgments

I am indebted to three of the great public institutions of New York City—the Municipal Archives, the Public Library, and Baruch College of the City University of New York—for providing resources and assistance that were indispensable in completing the book. In addition, Antoinette Georgiades, Moyee Huei-Lambert, and Anne-Marie Sheehan deserve special thanks for their great help in typing the manuscript.

Many people generously lent their time to recall details of events now thirty years old. I especially wish to thank former mayor Robert Wagner, William Shea, and former baseball commissioner Bowie Kuhn for their recollections of the events in New York. Rosalind Wyman, Kenneth Hahn, Roger Arneburgh, Dick Walsh, and Congressman Edward Roybal recalled the intricate details of the negotiations between the Dodgers and Los Angeles and the subsequent political and legal battles in that city.

Peter O'Malley graciously discussed the implications of the move in several meetings. Fred Claire, Sam Fernandez, Dave Van de Walker, and Vin Scully of the Dodgers took time from busy schedules to help clarify specific matters. Steve Brener, the Dodgers' publicity director, replied generously and promptly to repeated requests for information, and I am deeply grateful for his help.

Equally valuable (if less glamorous) sources of information were tapped for me in the Los Angeles County Archives by Susan Freedman, whose research greatly facilitated the completion of the book and significantly improved its quality.

I was sustained during visits to Los Angeles by my good friends Mike and Lana Sremba, whose accommodations rate five stars. Tim and Andy Ireland kindly extended their hospitality and encouragement to me during visits to Washington. Ron Yoshida, Roger Hannon, Dave Lockhart, and Nat Cipollina discussed many of my ideas with me and helped clarify some important points.

Sheldon Meyer, Joellyn Ausanka, Rachel Toor, and Sam Tanenhaus of Oxford University Press greatly improved the manuscript. Gail Ross provided valuable assistance, which is most appreciated.

I learned the essentials of writing from Peter Woll, my mentor and friend, at Brandeis University. His remarkable productivity remains inspiring, and his enthusiasm for writing, contagious.

Dan Fenn has provided valuable counsel and assistance for many years, and I am grateful for his friendship.

My wife Joyce shared the excitement and enjoyment of this project, and has enriched this effort as she does every aspect of our life together. The baby who will grace us in the spring has already put the book in its proper perspective. My nephew Rob served as proofreader and inspiration. Finally, my fascination with baseball and politics was nurtured by my father, and it is to him and to my mother, who endured countless dinner conversations about these subjects, that this book is dedicated.

New York   N.J.S.
September 1986

Contents

The Dodgers Move West

1

Brooklyn's Place in the New York Game

On September 24, 1957, Danny McDevitt of the Brooklyn Dodgers pitched a 2–0 shutout against the Pittsburgh Pirates, and thereby concluded one of the most colorful chapters of sports in America. After sixty-seven years as a National League franchise, the Dodgers would play no more games in Brooklyn.

The ending was rather pathetic. The Dodgers' move to Los Angeles had been assumed for months, but no official announcement had been made. The 6,702 fans who attended the finale in Ebbets Field—the team's home since 1913—were serenaded for nine innings with farewell dirges played by organist Gladys Goodding; but when the game ended it seemed nothing had changed. The public address announcer issued the customary injunction to stay off the field, and the ground crew performed its familiar tasks.[1] Some people took these rituals for signs that the team might stay after all. Even the players, dispersing at season's end, were uncertain where the Dodgers would play in 1958.[2] But on October 8, 1957, all doubts were dispelled by the announcement that the Dodgers were indeed moving to Los Angeles. Without fanfare or ceremony, Brooklyn lost the team that had figured so importantly in its history.

This book explains how local politics in New York and Los Angeles affected Walter F. O'Malley's decision to move the Brooklyn Dodgers to Los Angeles following the 1957 season and how government officials in both cities reacted to O'Malley's pursuit of land for the purpose of constructing a new stadium for his team. O'Malley stated repeatedly that he intended to construct a new ballpark with his own capital, but that he needed government assistance to acquire suitable land. New York officials were unwilling to furnish him with the site he wanted in

3

Brooklyn, at Atlantic and Flatbush avenues, while Los Angeles offered three hundred acres in Chavez Ravine, centrally located in the Los Angeles basin.

Other considerations influenced O'Malley's decision to move the Dodgers: his ability to persuade the New York Giants' owner, Horace Stoneham, to transfer his team to San Francisco; the advent of jet airline service, which removed the impediment of extended travel for other National League teams; and the migration of the middle class to the suburbs, which made fans reluctant to attend night games in the inner city. Yet of all the factors contributing to the Dodgers' decision, the political choices of the two cities were ultimately decisive. In the end it was more important to Los Angeles politicians to attract the Dodgers than it was for New York politicians to keep the team.

If the demise of baseball in Ebbets Field was pathetic, its inauguration had been triumphant. The first game was an exhibition between the Dodgers (or, as they were known at the time, the Superbas) and the New York Yankees. The lead story in the *Brooklyn Eagle* of April 5, 1913, exulted, "Twenty-five thousand hearts thumped with joy, twenty-five thousand pairs of feet pounded the concrete floor, and twenty-five thousand voices roared with delight—the day of days had at last arrived."[3] The day was a complete success as the Superbas defeated the Yankees, aided by a young outfielder, Casey Stengel, who hit a home run.

In 1913 Ebbets Field was the most modern and spacious of ballparks, constructed in slightly over a year at a cost of $750,000 in a part of the borough so desolate it was known, variously, as Pigtown, Goatville, Tin Can Alley, and Crow Hill.[4] During the ground-breaking ceremonies on March 5, 1912, the borough president, Alfred E. Steers, recounted how as a boy in 1870 he had spied through a knothole at the Capitoline Grounds and watched the Brooklyn Atlantics' famous victory over the Cincinnati Red Stockings.[5] The Atlantics were one of several teams that had played in Brooklyn since at least the 1850s.

A myth of amateurism attended baseball's development, in part because the first teams included wealthy members who formed gentlemen's clubs. Henry Chadwick, the first journalist to focus on baseball, further promoted the aura of amateurism by encouraging clubs to follow the gentlemanly traditions associated with cricket and rounders in his native England. Since Chadwick lived most of his life in Brooklyn, he doubtless was aware that his idealism was not always shared.

As David Voigt has written, the early history of baseball included tension between the upper-class amateurism of the gentlemen's clubs and the economic self-interest cum civic pride which spawned professionalism. These later factors triggered the development of several professional leagues after the Civil War.

Amateurism was doomed by the growing popularity of the sport. The first of the present major leagues, the National League, was formed in 1876. Its mission was not only to provide audiences with quality baseball, but also to uplift the moral standards of the sport. Sunday games were prohibited as were gambling and liquor consumption on the grounds. Brooklyn teams played high-caliber baseball but failed to be admitted into the league, in all likelihood because they were not suitably genteel. During these early years of major league play, Brooklyn teams were relegated to competing in minor circuits such as the Interstate League.[6] In 1881 the American Association was established; it offered serious competition to the National League and scandalized those who objected to Sunday games. Brooklyn qualified for the American Association in 1884 and capped its final year of membership in 1889 by winning the league pennant.

In 1883 the National League and the American Association reached an agreement that imposed reserve-clause controls on the players (that is, measures allowing clubs to renew expired contracts, thus binding players to one team) and also established the first postseason playoffs between league champions. A year later, however, the Union Association formed, and attracted players by refusing to recognize the reserve clause. That move proved to be the new league's undoing as its teams raided one another's rosters; only five of the original twelve clubs were able to finish the season.

The players' frustration with the lack of an economic alternative to the two leagues led to the creation in 1890 of the Players League, an eight-team circuit that included a franchise in Brooklyn. This league developed from the Brotherhood of Professional Ball Players, formed in 1885, and was capably led. Still, it folded after a year, the fans' interest having been saturated by the three leagues.

Meanwhile, the agreement between the National League and the American Association was strained by hard economic times, and in 1890 the National League invited the Brooklyn Trolley Dodgers, the Association's champions the preceding season, to switch leagues.[7] The team accepted, contributing to the demise of the Association, which folded after the 1891 campaign. Before the 1890 season, six Brooklyn players married and the team became known as the Bridegrooms. They remained champions, winning the National League pennant in 1890.

In the 1890s baseball continued to act in ways that contradicted latter-day misconceptions about the game's innocent past. A turn-of-the-century episode involving the Baltimore Orioles surpasses any modern intrigue and puts the lie to contemporary nostalgia for the age of sportsmen motivated purely by love of the game.

The affair began in January 1898 when Charley Byrne died and Charles Ebbets acquired the presidency and part-ownership of the Dodgers. By then the club had fallen on such hard times that by the end of the season Ebbets, who had started his career as a ticket taker and program vendor, was also serving as field manager, piloting the team to a tenth-place finish.[8] In a bold stroke, Ebbets cut a deal with Harry Von der Horst, owner of the Baltimore Orioles. Although the Orioles were the most formidable team of the time, with five future Hall of Famers, Baltimore fans no longer supported the club. Von der Horst was in a financial bind. As a remedy he sent Manager Ned Hanlon and star players Willie Keeler and Hughie Jennings to Brooklyn in exchange for a financial interest in the Bridegrooms. In addition to this personnel, Ebbets received a share in the Orioles, engaging in the kind of syndicate ownership common at the time. Managed by Hanlon and infused with Oriole talent, Brooklyn captured pennants in 1899 and 1900.

The National League's monopoly prevailed until 1901 when another of the small independent leagues of the era, the Western Association, headed by Ban Johnson, renamed itself the American League and stole some of the best players from the senior circuit. After first ignoring this upstart rival, the National League sought peace in 1902. The truce between the two, like the earlier one between the National League and the American Association, constrained business competition and ensured the owners' control over their players.

A loser in the resulting pact was Baltimore, which saw its team transferred to New York in 1903 as the National League Orioles became the American League Highlanders and later the Yankees, providing the entry into the New York market that the new league required for status and survival. To repair that loss, Hanlon conspired with Von der Horst to move Brooklyn's team to Baltimore, but the shift was thwarted by Ebbets's purchase of the remaining shares of the Dodgers and by his commitment to keep the team in Brooklyn. Had he not intervened, the history of major league baseball in Brooklyn might have been limited to the nineteenth century, as was the case in many cities.[9]

Hanlon continued to manage the club, but proved unable to win once his former Orioles retired and after 1902 Brooklyn floundered. The interest in baseball continued to grow in New York, however, especially with the arrival of John McGraw as manager of the Giants. McGraw was still another of the great Oriole players of the 1890s, and along with his former teammate Wilbert Robinson, who became a Giant coach, he brought the science of "inside baseball" to New York. Led by future Hall of Famers such as Christy Mathewson, Rube Marquard, and Joe McGinnity, the Giants won pennants in 1902, 1905, 1911, 1912, and 1913.

During this period Brooklyn seethed. In 1897 it lost its status as an autonomous city by being consolidated into greater New York, a move that meant compelling financial savings, but that also reduced Brooklyn to a rural enclave of sophisticated Manhattan. Many Brooklynites referred to consolidation as "The Great Mistake," and McGraw's success, combined with the Dodgers' futility, intensified the rivalry between the teams and the boroughs. Ebbets began to rectify the imbalance in 1912 by hiring Wilbert Robinson away from the Giants after Robinson had fallen out with McGraw. Ebbets also began developing players who would be stars in a few years, and he started construction of the stadium that would bear his name. Ebbets Field was just one of several new parks built about 1912, when a construction boom reflected the burgeoning popularity of baseball and the tremendous profits realized by owners. Comiskey Park and Wrigley Field in Chicago, Fenway Park in Boston, and Navin Field, now Tiger Stadium, in Detroit, still remain from that period.

The peace of 1902 between the major leagues was threatened in 1914 by a new competitor.[10] The Federal League consisted of nine franchises, including one in Brooklyn. Once again the reserve clause was ignored, and players in the established circuits used the threat of moving to a new team as leverage to win significant salary increases. Casey Stengel and Zack Wheat of the Dodgers battled Ebbets for pay hikes that were given grudgingly. The Federal League collapsed after the 1915 season, and many owners, including Ebbets, attempted to restore salaries to the level of "normal times." Connie Mack dismantled his brilliant Philadelphia Athletics team rather than increase its wages. The Federal League's team owners sued their counterparts on antitrust grounds in a case presided over by Judge Kenesaw Mountain Landis, who would become the first commissioner of baseball in 1920. Although most Federal League teams failed to survive even one year, the National and American leagues reached a settlement with them rather than risk a damaging decision on such issues as the legality of the reserve clause. David Voigt describes Landis's handling of the case as calculatedly dilatory, giving the owners time to negotiate a truce.[11]

Amid the chaos of the Federal League and World War I, Brooklyn mounted its first serious run at a National League pennant in the modern era. As the team clung to a narrow lead during the stretch drive of that year—1916—the *Brooklyn Eagle* divided its headlines between baseball and the battle of the Somme. The Superbas clinched the flag in the final week of the season, at last defeating the hated Giants in spite of that team's record twenty-six-game winning streak.

Brooklyn played its first World Series against the Boston Red Sox. The American presidential campaign was in its final month, with the incumbent Woodrow Wilson arguing that he had kept the country out of war. Technically this was true, but the championship began in Boston amid reports that German submarines were operating off Nantucket Island.[12] Boston won the Series four games to one. Red Sox pitching stifled the Superbas. Babe Ruth was hitless in five trips to the plate, but in the second game he inaugurated his World Series record of pitching twenty-nine and two-thirds scoreless innings. Prosperity undid Brooklyn. The team collapsed to a seventh-place finish in 1917 and McGraw's Giants won the pennant. After too many salary disputes for Ebbets's patience, Casey Stengel was traded at the end of the season, leaving Zack Wheat as the undisputed star of the team. Wheat led the team to its second pennant in 1920, another troubled year for the sport and the country.

In 1920, fundamental changes in the organizational structure of baseball were introduced, along with changes in the way the game was played. The key organizational change was that the three-member National Commission, which had governed baseball since 1903, was replaced by a single commissioner, Judge Landis, who had presided over the settlement between the National and American and Federal leagues in 1916.

Landis's initial task was dealing with the Black Sox scandal, in which eight Chicago White Sox were accused of throwing the 1919 World Series against Cincinnati. Although the players were acquitted in court of all charges of fraud, Landis banned them from the game for life and expunged their records from the official compilation of baseball statistics. These measures completely ignored the players' civil liberties, but effectively eliminated the suspicions of gambling that had tainted the game since the nineteenth century.

Equally important in restoring baseball's appeal was the emergence of Babe Ruth as a home run hitter and as a major figure of American culture. Traded from the Red Sox to the New York Yankees before the 1920 season, Ruth ushered in the demise of "inside baseball" tactics, offsetting the calculations of an entire game with one swing of the bat. That same season, Wilbert Robinson proved himself as capable as any manager in the game, guiding the Dodgers to the pennant with virtually the same players who had won in 1916.

During the final months of the 1920 season public attention was directed to a perceived threat to the country, as it had been four years earlier; but a foreign war was not the issue this time. "The Red Scare," with its hysterical fear of Marxism, dominated the national news. The *Brooklyn Eagle* focused on a trolley union strike in Brooklyn and on September 16, 1920, a bomb exploded near the Morgan bank on Wall Street, killing thirty-seven people.[13] Baseball coverage provided a bizarre forum for the expression of social concern. Thomas Rice, who covered the Superbas for the *Eagle*, mixed strident attacks on aliens, Trotskyites, and conscientious objectors to World War I into his stories. For instance, after Brooklyn appeared headed for the league championship, Rice wrote:

> If the Superbas were to lose the pennant now would be as big a surprise as it would be for a Greater New York police magistrate to show sufficient gumption and political courage to inflict heavy penalties upon the young thugs who, in the middle of every September for the past few years, have been allowed to mob decent people who chose to wear a straw hat longer than the period set by the gangs of juvenile mongrels, mostly of alien origin, whose assaults upon citizens have been treated by the magistrates as a merry jest to the vast encouragement of all delinquents. The failure of the Superbas to win out now would bring fits upon Flatbush and environs, just as the timidity of the magistrates in suppressing the hat demolishing mobs is sure to lead to homicides.[14]

The Republic survived, and the Dodgers prevailed. They played Cleveland in the World Series, but again won only one game in a Series remembered now for two feats accomplished by Cleveland: the first grand slam home run and the only unassisted triple play in World Series history.

From 1921 to 1924 the New York Giants reasserted themselves with four consecutive National League pennants. The Dodgers came close only in 1924, losing by a game and a half. For the rest of Robinson's managerial tenure, which ended after the 1931 season, the team was mediocre. It was at this time, when success proved elusive, that the club developed its comic reputation. The Dodgers were known as the "Daffyness Boys," and Brooklyn itself acquired a colorful image while admiration and acclaim went to Ruth and the Yankees. The Dodgers of that era were epitomized by their right fielder, Babe Herman, an excellent hitter who batted .324 in a career spanning thirteen seasons. In 1920 he hit .393. Twenty-five years later, at age forty-two, he returned to play for the team, which had been decimated by the World War II draft. Even though he had been out of baseball for eight years, Herman still hit .265 as a pinch hitter. Despite those impressive credentials, Herman is best remembered for his wild base-running and his inept fielding. He swore he had never been hit on the head by a fly ball—the neck and shoulders didn't count. He hit perhaps the most remarkable triple in the history of the game; it resulted in three Dodgers, including Herman, occupying third base. Dodger pitchers of the 1920s included Burleigh Grimes, among the last of the authorized spitball pitchers, Dazzy Vance, whose tremendous fastball was aided by a distractingly shredded red undershirt, and Van Lingle Mungo, whose name alone added panache to the team.

The Dodgers' image was one of zany characters loyally supported by a borough that embraced cartoonist Willard Mullin's image of the bum as clown. But in reality turmoil plagued the front office, and until 1939 the team was close to receivership. Attendance fell as fans demonstrated that they would flock to Ebbets Field only if the team won.[15]

The 1920s and 30s were difficult years on the field and off. In the spring of 1925, Charley Ebbets suddenly died, leaving his half ownership scattered among his surviving family. The other half of the club belonged to the McKeever brothers, Steve and Ed, whose construction company had come to Ebbets's financial rescue during the building of his stadium. Ed McKeever replaced Ebbets as president of the franchise, but during the burial service for Ebbets he contracted pneumonia and a week later was dead himself. Steve McKeever and the other team directors chose Wilbert Robinson to fill the post of president. Robinson, supported by Ebbets's heirs, battled endlessly over the operation of the Dodgers, and the strife damaged the club. In 1930 a truce was reached securing Robinson's place as manager but forcing him out of the presidency.

By 1932, Robinson lacked the support even to remain as manager, and after eighteen years was let go. Throughout the 1930s, the team floundered on the field under the leadership of Max Carey, Casey Stengel, and Burleigh Grimes, and the organization, rocked by the Depression, was reduced nearly to bankruptcy, as the quarreling owners were unable to formulate a solution to these woes.

Meanwhile, the Giants won pennants in 1933, 1936, and 1937 under manager Bill Terry, who succeeded McGraw when he retired in 1931; but baseball during the 1920s and 1930s meant the Yankees of Ruth, Lou Gehrig, and Joe DiMaggio. No team has ever dominated the sport so completely. The Yankees became the image of New York glamour, while the Dodgers, along with Brooklyn, became a comic foil. To Red Barber, who brought the perspective of a Southerner to his job as the Dodgers' radio announcer, Brooklyn had a collective inferiority complex, which caused it to embrace the caricature of the bum and to utter the familiar cry, "Wait till next year!"[16]

In 1937, desperate about the club's condition, the Dodgers' owners turned to league president Ford Frick for advice. Frick recommended they hire Leland Stanford MacPhail. In his book, *The Brooklyn Dodgers*, Frank Graham recounts Larry MacPhail's introduction to the Dodgers' board of directors. When offered the position of president, MacPhail replied

> "Sure I'll take the job—if you lay the kind of dough I want on the line for me, give me a free hand, and fix it up with the bank so that when I want some real money for operating purposes I can walk in there and get it."
>
> They realized it would take a lot of money. How much did he want, for instance? When he told them, they paled. And what was his idea of operating expenses?
>
> "We won't put any limit on that," he said.
>
> They grew paler. He looked at his watch, looked at them and said, "Well, make up your minds. If I can't do any business here, I know where I can. And if I'm losing here, I'd like to know about it, because I've got to be on that three o'clock train for Washington." He smote the top of the desk sharply with the flat of his hand. "Well," he said brusquely, "what about it?"[17]

The directors accepted his terms, and hired the forty-eight-year-old executive. MacPhail had pursued a brilliant but erratic career that included an appointment to the Naval Academy at Annapolis, which he turned down, a stint at the University of Michigan Law School, and service in World War I, in which he nearly succeeded in kidnaping Kaiser Wilhelm. As vice president of the Cincinnati Redlegs, MacPhail had introduced night baseball and the first radio broadcasts of games. Before the start of the 1938 season he announced his intention to restore the Dodgers to a competitive level, and to develop their farm system to provide for the future. He established the seriousness of his demands by informing George V. McLaughlin of the Brooklyn Trust Company that he needed $50,000 to acquire first baseman Dolf Camilli from the Philadelphia Phillies. The sum was unheard of in those days, at least in those offices. McLaughlin agreed, however, and the Dodgers were on their way.

While securing the money for Camilli, MacPhail also informed McLaughlin that Ebbets Field was in need of renovation, and that $200,000 would do the trick. This amount was forthcoming, too, in part because MacPhail had insisted his contract specify that "He will have full and complete authority over the operations of the club. . . ."[18]

The stadium was repainted, given new seats, renovated dugouts and clubhouses, and the field was groomed to eliminate rocks and divots that had plagued infielders for years. MacPhail also hired genuine ushers to assist fans in finding their seats, replacing the goons who traditionally had preyed on unwary patrons who were as likely to be mugged as aided. A new press box with an adjacent lounge, complete with bar, improved public relations. And a prevailing agreement among the Dodgers, Yankees, and Giants not to broadcast their games was immediately scotched by MacPhail, who lured his old announcer, Red Barber, away from Cincinnati.

A more dramatic innovation was the introduction of night baseball to Ebbets Field. MacPhail had originally said there was no need for night scheduling in Brooklyn, but he changed his mind by the start of the 1938 season. The lights were ready for a game against Cincinnati on June 15, when Johnny Vander Meer, fresh from a no-hitter against the Boston Braves, repeated the feat against the Dodgers, the only consecutive no-hitters in the history of the game. Perhaps no team has ever been transformed so drastically in so little time—only a few months. When Steve McKeever died on March 12, 1938, it symbolized the end of an organization unable to meet the challenges of the modern era.

MacPhail, equally attentive to the importance of fielding a competitive team and to the necessity of promoting the club in the entertainment market, brilliantly established the basis for the Dodgers' subsequent success. He hired Babe Ruth as a coach in order to treat fans to a glimpse of the Babe taking batting practice. Other promotional stunts anticipated the antics of Bill Veeck and Charlie Finley. The team nevertheless stumbled to a seventh-place finish in 1938. The following season, shortstop Leo Durocher replaced Burleigh Grimes as manager and led the Dodgers to a third-place finish.

The talent improved in 1940, with the additions of Pee Wee Reese and Pete Reiser from the Dodger farm system and the acquisition from the Cardinals of Joe Medwick. The St. Louis Gashouse spirit affected the Dodgers, who made a run at the pennant, finishing in second place, twelve games behind Cincinnati. The Dodgers had been transformed from doormats into an exciting team that could be expected to challenge for the flag. The further addition of Dixie Walker, eventually known in Brooklyn as "the People's Cherce," and of pitcher Kirby Higbe and catcher Mickey Owen, added the elements needed for a pennant. The 1941 season brought Brooklyn its first championship in twenty years. The team won one hundred games, three more than St. Louis, as Camilli led the league in home runs and Reiser won the batting title. Higbe and Whitlow Wyatt tied for the league lead in wins with twenty-two apiece. True to his word, MacPhail had restored the team to a competitive footing and laid the groundwork for future prosperity.

The 1941 World Series was the first between the Dodgers and the

Yankees. They split the opening two games in Yankee Stadium, the Dodgers rallying from a 2–0 deficit in the second contest. One of the bitterest moments in the memory of Brooklyn fans occurred in the fourth game as the Dodgers held a 4–3 lead with two out in the ninth inning and were on the verge of tying the Series again. Hugh Casey threw a third strike past Tommy Henrich, but the ball also eluded the catcher, Mickey Owen, and Henrich took first. DiMaggio then singled. A double by Charlie Keller, a walk, and another double scored four runs and gave the Yankees a three-game-to-one lead. When New York closed out the Series with a 3–1 win in the fifth game, it was almost anticlimactic. Thereafter, World Series meetings between the Dodgers and Yankees gained in intensity as Brooklyn struggled to overcome the fate that seemed to doom them to a second-rate status within New York City.

The 1942 season was a turning point for the Dodgers. The team had yet to be hit by the military draft and racked up an unprecedented 104 victories. Unfortunately, the Cardinals won 2 more. By season's end MacPhail's time with the Dodgers came to its conclusion. The club's success, which included attendance figures of more than one million per year, did not calm the board of directors, who chafed under MacPhail's continued insistence on investing revenues in the team's future. Expenditures for players irked the owners, who had yet to receive significant dividends from their outlays. MacPhail also had a stormy relationship with Durocher, league officials, umpires, and the press. Following the pattern he had established in previous ventures, MacPhail decided it was time to move on. He bade a tearful farewell to Brooklyn and turned his talents to the war effort.

From today's perspective, one can see that, for all the tumult his personality caused, MacPhail made the Dodgers a modern organization. He rescued the club from bankruptcy, invested in its future, promoted the players, and upgraded Ebbets Field. He helped transform the farm system into a model for other teams, and he imposed an expectation of success that has made the Dodgers serious competitors ever since. Despite all the changes that have since occurred, the roots of the Dodgers' organization and its impressive accomplishments originated in the era of Larry MacPhail.

When MacPhail left, the directors hired Branch Rickey, who had made the St. Louis Cardinals the league power. Rickey had developed the game's best farm system with St. Louis, and his tenure with the Dodgers produced the team that won six pennants from 1947 through 1956. Those Dodger teams, heralded as "the Boys of Summer" in Roger Kahn's wonderful book of that title, were by far the most successful in Brooklyn's history.[19] Even their defeats were legendary, especially the loss in the 1951 playoffs to Bobby Thomson and the Giants. One half the team's regular starters—Duke Snider, Jackie Robinson, Roy Campanella, and Pee Wee Reese—were later inducted into the Hall of Fame. The World Series losses continued, but the outcomes were closer, and in 1952 and 1953 they were virtually the equal of the Yankees. In 1955 the Dodgers finally proved their character. After dropping the first two games of the Series, they became the first team to recover from such a deficit, winning the Series against the Yankees in seven games for the only world championship in Brooklyn's history.

Baseball in New York during the postwar years was glorious.[20] The Yankees won all but two American League flags between 1947 and 1958 and won an unprecedented five consecutive world championships from 1949 to 1953. Casey Stengel, who had failed to make the Dodgers into winners, was hailed as a genius, a worthy successor to Miller Huggins and Joe McCarthy. The Giants fared less spectacularly during those years, but won pennants in 1951 and 1954. The victory in 1951 was especially satisfying, since they trailed the Dodgers by more than thirteen games as late as August 12. Bobby Thomson's home run in the bottom of the ninth inning of the decisive playoff game is still perhaps the most dramatic moment in baseball history. In the 1954 World Series, the Giants swept the Cleveland Indians on the strength of Willie Mays's fielding and the pinch-hit home runs of Dusty Rhodes.

The rivalry among the three teams was intense, and fan loyalties were strained by the owners' machinations. Stengel had been one of the most beloved characters in the Dodgers' history, yet he now managed the hated Yankees. Leo Durocher, who managed the Dodgers back to greatness in the 1940s was suspended for the 1947 season because of his alleged association with gamblers. Those charges were leveled by Larry MacPhail, back from the war and now president and part-owner of the Yankees. When Durocher returned to baseball in 1948, he briefly piloted the Dodgers before going across town to manage the Giants. The success of the team led to fierce debates on the relative merits of Mays, Mantle, and Snider as centerfielders, Berra and Campanella as catchers, Reese and Rizzuto as shortstops. Baseball had come full circle—back to Chadwick's day, when it was New York's game—and fan support was at its peak.

Even in that exciting time, the attachment between the Dodgers and Brooklyn was a special case. What constituted the nature of the relationship is more a question of romance than of science: the quality of the Dodger teams, Brooklyn's large population, and the concentration of news media in New York undoubtedly contributed to the team's mystique. But they don't provide the full explanation. The Dodgers may have been so important to Brooklyn because they symbolized the borough's aspiration to escape a humiliating burlesque role. For all the community's economic and cultural achievements, its consolidation into New York had made it a kind of comic foil to Manhattan. The other boroughs may also have smarted as a result of their second-class status, but as the largest borough in the city, Brooklyn's diminished role required the sharpest adjustment.

The concentration of press coverage in New York put the Dodgers under a brighter spotlight than they would have received in other cities, but again Brooklyn was outside the first circle. In his biography of Red Smith, Ira Berkow describes the sportswriter's frustration after missing an opportunity to move from Philadelphia to a New York paper. "Smith was very disappointed. Another time he received an offer from the *Brooklyn Eagle*. 'But I decided,' recalled Smith, 'that Brooklyn was farther from New York than Philadelphia was.' "[21] This attitude betrayed a common perception of Brooklyn as a kind of province, an interesting and important place, but not New York. According to William Riordan's biography of Tammany Hall operative George Washington Plunkett, Brooklyn residents were dismissed as hayseeds. Many residents of Brooklyn saw the Dodgers as a means of establishing their rightful place in the larger city. The springs, summers, and autumns of promise, perennially unfulfilled, bound Brooklyn and the Dodgers in a union of frustration that intensified with each missed opportunity. The first missed opportunity was Mickey Owens's passed ball. The disappointment of the partisans in Ebbets Field who watched victory unravel in the ninth inning was conveyed by the *Brooklyn Eagle:*

> I don't want to talk about it, a mournful-looking father told his small son, and the lack of conversation in the jammed aisles and runways indicated clearly that he wasn't the only one who considered the debacle too bitter and rankling for words.
>
> A pretty girl in the upper tier behind first base who had covered her eyes when Keller's wallop headed for the right-field wall, stared as if hypnotized at a bibulous neighbor who had tippled too lustily and fallen into a sound sleep in the seventh inning.
>
> "Gee," she said, somewhat enviously, "He must still believe we're ahead, 4 to 3."[22]

A rematch with the Yankees was deferred until 1949 in a series that produced two memorable moments for Brooklyn fans: the winning ninth inning-rally that thwarted Floyd Bevens's no-hitter in the fifth game, and in the sixth, Al Gionfriddo's catch of Joe DiMaggio's drive, preserving a Dodger victory.

In certain respects, the Yankees mirrored the place of Manhattan in New York City government, the first among equals. Manhattan and the Yankees both dominated their respective spheres in a manner that conveyed arrogance and inevitability. The Yankees attracted fans from the entire metropolitan area to their ballpark in the South Bronx, which bordered the Harlem River, less than a mile from the Giants' park, the Polo Grounds, located in northeastern Manhattan. Of all the boroughs, Brooklyn alone identified closely with just one of the city's major league teams. When the Dodgers won a championship, it symbolically established Brooklyn's preeminence within the city.

The Dodgers spent two agonizing years trying to recapture the National League title. In 1950, the Phillies lost a nine-game lead over the Dodgers as the Whiz Kids took the championship on the final game of the season with a win over Brooklyn. Cal Abrams of the Dodgers was thrown out at the plate in the ninth inning trying to score what would have been the pennant-tying run, and Dick Sisler homered for Philadelphia to give that city its first title in thirty-five years. In his post-mortem in the *Eagle*, columnist Tommy Holmes noted a curious feature of Brooklyn fans. "I think I'd feel even more sorry for our town if the fans hadn't waited until the very last game before completely filling Ebbets Field for the only time this year."[23]

Holmes noted that the year had been far from a financial success for the Dodgers, and he proposed that money was one reason why Branch Rickey was selling his shares of the Dodgers. Holmes didn't mention the intra-office warfare between Rickey and Walter O'Malley, but he recognized a critical financial factor that offset the intensity of the fans' devotion.

If 1950 was disappointing, the following year was disastrous. A thirteen-and-a-half game lead in August dwindled as the Giants played torridly in the final part of the season. When Bobby Thomson delivered his legendary home run, Holmes spoke for many fans:

> Don't make mine vanilla on this round. I'll settle for cyanide and there must be thousands of people in our fair and unhappy town who feel about the same, as the Giants moved into action against the Yankees in the first game of the World Series today.
> This really was a rough one if you're the type to grow emotional over the game of baseball and are a fan of the Dodger persuasion.
> But it seems to be rough, year after year.[24]

Holmes captured the frustration of a community in possession of an excellent team—one of the best of all times—that continued to fall just short of its potential and of the hopes of its fans.

The *Eagle* also recorded Brooklyn's reaction to Thomson's home run, including one besotted conversation in a Myrtle Avenue bar, where two elderly gentlemen fantasized their way through four illusory victories over the Yankees.

"Well," said one of the old men, "baseball isn't everything."
"It isn't?" asked his equally old friend. "What else?"
"There's basketball," he replied.
"Not anymore," said his friend.
"There's women."
"At our age?" said his friend.
"Well, there's always beer."
"Yes," said the other old man, "there's always beer."[25]

When the years of disappointment finally ended in the championship season of 1955, the borough-wide elation revealed how frustrating the earlier defeats had been to Brooklyn. A fan interviewed by Peter Golenbock, author of *Bums,* an oral history of the Brooklyn Dodgers, recalled that he had been a Marine corporal stationed in Maryland when the Dodgers finally beat the Yankees. "I called my sister on the phone, and here I was in the brig guarding prisoners, and I said 'Ronnie, what is it like in Brooklyn?' And she said, 'Listen to this,' and she stuck the phone out the window, and I could hear the roar of the crowd, the screaming. And I wanted so badly to be there, I just started to cry, and the prisoners, a couple of sailors, they started laughing when they saw me. Here I am, a big marine, and I'm crying, tears rolling down my face. Those were great days. The greatest in the world."[26]

Donald Honig told Golenbock, "It was almost as though you were finding your manhood at stake. You rooted for this team, and every October it would die. And when you put the win in the context of a small neighborhood, where your personal relationships were very tight, it became even more important. The Yankee fans and Giant fans were always ribbing you, saying that you choked. Dodger fans very well knew the sentiments of the mythical Man on the Street. He knew that it was said that the Dodgers choked every October, and what the hell were you going to say? They did lose every October.

"I know all this sounds so superficial—your team won, so what? A nonfan could not conceive of how important it was. But baseball was an important factor in my neighborhood. For twelve months a year, it was baseball and little else. Everyone knew who you rooted for before they knew your religion or nationality. It was important to you that your team did well. Baseball was very important, and rooting for the Brooklyn Dodgers was something special."[27]

In their emotional intensity, the Dodgers' fans often forgot the pressures threatening the team's continued presence in Brooklyn. When the prospect of losing the Dodgers became a public issue, residents wrote to Mayor Robert Wagner. Some of their arguments raised basic questions about the role of government, but many were simple appeals to keep the team in Brooklyn. One such plea, from a Mrs. Tyree Smith, began, "I am writting [sic] this letter as a loyal Brooklynite, New Yorker and Democrat. (I am putting them all together because they go together like coffee and cream with sugar). . . . If we lose the Dodgers and the Giants because you think they are bluffing, I will give you something to think about."[28] Another woman wrote: "You won't get my vote in the coming election, if you do decide to run again, if the Sports Center for the Dodgers isn't built so they will stay in Brooklyn." Still another voter, Mr. R. Cucco, put the matter more directly. "I am a man of very few words so I will come straight to the point. I voted for you. I pay your salary. I WANT THE DODGERS IN BROOKLYN. I don't want any excuses from you or any of your men there at City Hall. I WANT THE DODGERS IN BROOKLYN, and you can do it by building that Sports Center. You had better get it built or you'll not get a vote from me."

Even theology and sociology were invoked to influence Wagner's decisions. Gloria Cerrato wrote, "I am a young girl of 16 and enjoy baseball to great extents [sic]. If the Dodgers move to Los Angeles I will no longer enjoy this right given to me by my Creator. Please keep them here. Baseball keeps a lot of us teen-agers off the streets and prevents J[uvenile] D[elinquency].

"Let's keep the Bums in Brooklyn forever and ever, by building them their stadium. It won't only be for them. Remember."

Another letter took a more dispassionate approach: "I cannot impress upon you too much how important it is to keep the Dodgers in Brooklyn. It keeps the children off the streets during the day, it gives them someone to look up to, someone to imitate. Instead of acting like 'tough guys,' they try to imitate Duke Snider, Pee Wee Reese, Roy Campanella, etc. It also gives them a feeling for 'fair play.' And the Dodgers, being composed of Negroes, Spanish, and Whites, are a good example of how good you can get if everyone works together regardless of race or color."

The Dodgers were more than a business, more even than a sports franchise. They represented a cultural totem, a tangible symbol of the community and its values. Baseball and the Dodgers were a pastime which lured young people away from the inducement of crime and indolence. During the summer months when young people were not constrained by school, the Dodgers served as a kind of babysitter. From our perspective, these people's trust in baseball's saving grace may strike us as naive; but our culture had barely been introduced to rock music let alone the other distractions that would follow. Brooklyn was undergoing many changes, and the Dodgers were one of the few means for getting one's bearings.

The Dodgers also played a role in improving race relations. Like the Giants, the Dodgers aggressively pursued black athletes as an untapped market of talent. This might seem a natural occurrence in a multiracial city such as New York, but the Yankees didn't sign a black player until Elston Howard joined the team in 1955.

This team would have attracted a following in Brooklyn solely on the basis of its many exasperating failures to win championships, but much more than their performances on the field bound the Dodgers to Brooklyn. The itch to escape the tag of bumpkin, the possibility of racial harmony, and, perhaps most importantly, the futile hope that borough life would return to familiar patterns—all these sentiments bound the Dodgers to Brooklyn, and made their presence essential.

Even as the Dodgers and Brooklyn carried on their romance, the business of baseball was unobtrusively adapting to new conditions. The impact of television, the mass exodus to the suburbs, the popularity of the automobile, and population shifts to the South and West affected the financial stability of most teams. No one had a keener sense of these changes than Walter O'Malley, for whom sentimentality about the Dodgers' past would not be sufficient to keep the team in Brooklyn.

2

Postwar Challenges and the Beginning of the O'Malley Era

The 1940s were the first period of sustained success for the Dodgers in thirty years. The team won pennants in 1941, 1947, and 1949, and tied St. Louis for first place in 1946 before losing in a playoff. Attendance reached an all-time high in 1947 when 1,807,576 fans paid to see the Bums.[1] This era represented the team's emergence as a durable power in the National League. One would suppose that this turnaround was directed by a stable coterie of owners and executives who pursued a coherent strategy for fielding strong teams and pulling people into the stadium. In truth, the club was in turmoil, and the team's operations were guided by three presidents who displayed brilliant business skills and a gift for winning but who were incompatible in temperament and in their choice of tactics.

Larry MacPhail held the briefest tenure, from 1939 to 1941. His innovations saved a moribund franchise. When he resigned, the board of directors—the Brooklyn Trust Company—was determined to see its investment remain in capable hands. They seized the opportunity presented in St. Louis by a clash between the Cardinals' president, Branch Rickey, and their owner, Sam Breadon. Rickey accepted the offer to become the president of the Dodgers, assuming the post after the 1942 season. The move afforded him an unfettered opportunity to run a ball club, and it also reunited him with his son, Branch Jr., who had served in St. Louis before leaving to run the Dodgers' farm system.

In his own way, Rickey was every bit as daring and ambitious as MacPhail, although his manner was almost totally different. Rickey was born in 1881, and moved two years later with his family to a farm in Duck Run, Ohio.[2] From his vantage point, Cincinnati must have seemed the epitome of urban life, yet in the New York papers of that period the Red Stockings were derided for representing "Porkopolis."

Rickey's life was a reflection of traditional entrepreneurial values. He worked hard with his family on their farm, and walked miles to a one-room schoolhouse. He married his childhood sweetheart, and all his family relationships were imbued with the lessons found in the Bible. To describe such origins as humble is patronizing, since Rickey grew up with a clear sense of purpose, a strong ego, and an abundance of ambition. These qualities enabled him to complete his college education at Ohio Wesleyan and to earn a law degree from the University of Michigan. A good athlete, he played baseball and football, and in his mid-twenties he played major league baseball. His career was mediocre, but his understanding of the game was thorough.[3]

He also coached. At the age of twenty-one, when he was piloting Ohio Wesleyan's baseball team, he angrily championed the right of his black first baseman, Charles Thomas, to compete against any team and to stay with his teammates in any hotel during road trips. These encounters with Jim Crow laws in the Midwest left a lifelong impression on Rickey, who came to see coaching and sports as a means of molding character and instilling in his charges the moral values at the core of his life. At the same time, he was completely dedicated to winning, and welcomed the material gains that accrued from success on the field. Rickey would eventually use his position in Brooklyn to topple the color barrier that had long hampered the game. Rickey's concerns for equal opportunity stemmed both from the deep religious beliefs that guided all his activities, and from his keen business sense—he could not bear to see resources squandered.

In 1911 Rickey opened a law firm in Boise, Idaho with some former classmates. The partners anticipated a boom in the community as soon as the railroad opened the area to wider commerce, but no railroad materialized, and Rickey, now thirty years old, returned to Ann Arbor, Michigan, and resumed coaching. In 1913 he assumed his first executive position in the major leagues with the St. Louis Browns. In this capacity, he began to develop the farm system that would later be so critical to the success of the Cardinals and the Dodgers. Rickey built up the Browns' farm system with college players he had observed in his coaching days at the University of Michigan, but the parent team continued to flounder. In desperation, their new owner, Robert Hedges, hired Rickey as field manager near the end of the 1913 season. Rickey began his first full season as manager by introducing techniques that revolutionized spring training. Sliding pits, running drills, and handball games were among the innovations he anticipated would gain an advantage for his team. He tirelessly lectured his players on esoteric tactics, and threw in moral instruction as well.

When the economic uncertainty created by World War I and the rivalry of the Federal League led Hedges to sell the Browns, Rickey clashed with the new owner, then grabbed the opportunity to assume the presidency of the bankrupt Cardinals in 1916.[4] His career was interrupted, however, by the war. Although well past conscription age, Rickey enlisted in the army's chemical warfare unit, with the rank of major. He went to France with other major leaguers and drilled the troops in techniques for avoiding mustard gas. He had survived tuberculosis fifteen years earlier, so his voluntary exposure to a potential pulmonary hazard was genuinely heroic.

After the war, Rickey concentrated on developing the Cardinals' farm system, although between 1919 and 1924 he managed the Cardinals, without much success. He achieved winning records only from 1921 to 1923, and his teams never finished better than third.[5]

Even as a manager, Rickey refused to participate in Sunday games, entrusting the team to a substitute. For many people, this practice exemplified Rickey's sanctimonious piety. There is probably no truth to the often-repeated explanation that Rickey had promised his mother he wouldn't play on Sunday. As Red Smith later wrote, "She never asked him not to play on Sunday and he never promised he wouldn't. He just didn't."[6] In any case, Rickey's disapproval of Sunday baseball was not anomalous. The National League included a ban on Sunday play when it was formed in 1876.[7] And, even after the league capitulated, many cities upheld the ban. Even New York, for all its cosmopolitanism, enforced the Sunday ban until Al Smith was elected governor in 1918. St. Louis and Cincinnati, on the other hand, lifted restrictions in 1902. Boston obeyed the prohibition until 1929, and in Pennsylvania legislation imposed a curfew on Sunday games that persisted into the 1950s.

Rickey attracted disdain not because of his scruples, which were widely shared, but because his personality combined deep attachment to traditional values with keen business acumen. As Red Smith described the attitude of Rickey's foes, "He was pictured as a hypocrite who would outslick you in a trade on Saturday but considered Sunday baseball immoral."[8] As others put it, he would miss the game on Sunday, but greedily count the receipts on Monday. Those who preferred baseball to be strictly either a pastime or a business enterprise were invariably confounded by his practices.

Significant opposition to Rickey's development of the St. Louis farm system came from another figure who also combined strong moral and business precepts, Kenesaw Mountain Landis, who cleansed baseball of the corruption of gamblers, and who was Rickey's equal in moral fervor, business sense, and ego.[9] To Landis, farm systems reintroduced syndicate ownership, which smacked of corruption and threatened to poison the wellspring of the game: independent minor league franchises supported by local communities. If one of these teams could suddenly be stripped of its best players on the whim of a major league executive, what would become of the community loyalty on which the franchise depended for survival? Landis regarded farm systems as ruinous to the lower leagues and as a threat to the entire game.

With equal fervor Rickey contended that minor league teams could survive financially only if allied with wealthier major league patrons.[10] Moreover, if the major league teams could not develop and draw on talent, then the pennant races would always be dominated by the same few teams who could pay the largest salaries as they raided minor league rosters for the best players available. Competition would lag, and fan interest wane, as had happened in Baltimore in the 1890s. This conflict between Landis and Rickey anticipated the bitter dispute that emerged after World War II over the impact of major league expansion on the Pacific Coast League.

David Voigt's analysis of the game's history credits Rickey with perceiving some of the country's major cultural shifts. The growth of mass communications, for instance, especially radio and newsreels, focused people's attention away from their own communities toward glamorous urban centers, specifically New York. Even in the 1920s, often described as the Golden Age of Sport, mass communication elevated great sports figures, such as Babe Ruth, Jack Dempsey, Bobby Jones, and Bill Tilden, into national heroes known to audiences who might otherwise have been only dimly aware of their achievements. In this context, Landis's outlook was anachronistic. Just as the fabled amateurism of the game's earlier years was doomed by baseball's increased popularity, so the shift to mass culture precluded serious fan support of independent teams in small communities. Thus some minor league teams failed in the economic turmoil of World War I, and others were threatened by the Depression. Eventually even those minor league operators who decried Rickey's farm system eventually embraced it as their only means of survival.

Under Rickey's guidance the Cardinals prospered so well that they contended with the Giants for dominance in the league. They won their first pennant in 1926, and defeated the Yankees in seven games in the World Series. They won five more pennants and three world championships before Rickey left for Brooklyn after the 1942 season. Rickey's personality ill suited him for working with other ambitious men. Larry MacPhail had been brought into the St. Louis organization by Rickey but left for Cincinnati and greater authority. And, in spite of his success, Rickey irritated the Cardinals' owner Sam Breadon, their disagreements climaxing when Rickey opposed radio sponsorship for the team by a brewery. Breadon suggested that Rickey look for other opportunities and, when MacPhail left the Dodgers, Rickey made the move to Brooklyn.

Since Rickey's departure from the Cardinals ultimately hinged on his moral objections to alcohol, it is especially intriguing that his legacy to Cardinal baseball was the Gas House Gang, some of the least decorous—and most talented—characters ever to play the game. Rickey's first great Cardinal player was Rogers Hornsby, perhaps the greatest hitter in the history of the National League but also renowned for his miserable disposition.[11] Pepper Martin, Ducky Medwick, and Dizzy Dean thoroughly matched the colorful Dodgers and were far better at winning ballgames. While the Yankees won with mechanical proficiency, the Cardinals triumphed with a brawling aggressive style that matched Rickey's ideal of how the game should be played. He displayed a forgiving attitude toward the Cardinals' off-field adventures, but didn't hesitate to trade players when their skills begin to diminish. The key to Rickey's success was his ability to integrate the disparate elements of his personality and to nurture players who mirrored his own pragmatic daring.

Rickey's initial project with the Dodgers was to lay the foundation for bringing blacks into the game. In January 1943 he met with the president of the Brooklyn Trust Company, George McLaughlin, to determine his reaction to the prospect of an integrated team.[12] Rickey couched his arguments in business terms: the Dodger team that won the pennant in 1941 would be decimated by the draft for World War II; other teams would retrench during the war due to the strain on financial resources; accordingly, the Dodgers should seize the opportunity to secure the best available talent while their competitors were idle, and should consider using black players. McLaughlin, and at a later point his partners, agreed, their approval based not on moral sentiments but on economic self-interest. McLaughlin put the matter bluntly to Rickey: "If you're doing this to help the ball club, go ahead. But if you're doing it for the emancipation of the Negro, then forget it."[13]

The Dodgers' hiring of Jackie Robinson has been widely chronicled.[14] He had been an All-American football player at UCLA, where he also was on the track team. In 1942, at the age of twenty-three, Robinson requested and received a tryout with the Chicago White Sox at their training camp in Pasadena, California, joined by Negro League pitcher Nate Moreland. Robinson impressed the White Sox manager, but no further action was taken by the club. In April 1945, Robinson auditioned for another major league team, the Boston Red Sox. Again he performed impressively, but again a major league team feared to challenge the bias against integrated ballclubs.

World War II gave many blacks the opportunity to distinguish themselves in the military, and enabled others to leave the isolation of the South to secure employment in war-related industries. Moreover, the vicious racial doctrines of Nazi Germany forced America to examine its own racial attitudes. With exquisite preparation, Rickey signed Robinson to a Dodger contract in 1946 and, after a year with the Triple A franchise in Montreal, Robinson joined the Dodgers for the 1947 season. His arrival created friction with some of his teammates, a few of whom asked to be traded, and his first few years in the league included ugly taunts from opposing players.

Throughout this ordeal, Robinson remained outwardly calm, conforming to Rickey's injunction that he not fight back, regardless of the provocation. In the face of these pressures, Robinson was so effective that he won the major leagues' first Rookie of the Year title in 1947, compiling a .297 batting average and leading the league with twenty-nine stolen bases. Although his base-stealing totals are sparse compared with those of some later stars, his presence on base unnerved pitchers as perhaps no other player had since Ty Cobb.

In 1949 Robinson led the league in hitting with a .342 average, and was chosen Most Valuable Player. During the 1950s, he continued to hit over .300 and to play a critical role on the great Dodger team. He was probably their fiercest competitor, and his aggressive personality further spiced the rivalries in New York baseball during that period. Moving beyond his initial stoicism, Robinson broadened his attack on racial injustice. He attacked the slow pace of integration by many major league teams, and he spoke against the segregated conditions in hotels and restaurants that racially divided teams in many of the southern and border states. His militancy drew complaints from those who favored a more leisurely redress of grievances. After the 1956 season, the Dodgers tried to trade Robinson to the New York Giants, a proposal as remarkable as Brooklyn's acquisition of Sal Maglie a few months earlier. The transaction never materialized as Robinson announced his retirement from the game. In 1962 he was inducted into the Hall of Fame.

Robinson was no mere token; he was soon joined by other black players, including Roy Campanella, Don Newcombe, Joe Black, and Jim Gilliam. Once again, Rickey had acted boldly on the basis of an accurate perception of where America was headed. Other owners were either hostile to blacks or dilatory in their efforts to hire them; but Rickey again employed his farm system to groom black athletes who would someday help the team.

Landis, who remained commissioner until his death in 1944, again overlooked the signs of change, remaining, at best, ambivalent about blacks joining the major leagues; some evidence suggests he plotted against the move.[15] In a 1946 report to the new commissioner, Happy Chandler, the major league clubs voted 15–1 against admitting blacks to professional baseball except in the segregated Negro leagues.[16] Of course many beside Rickey were agitating for integration. Political figures, such as New York mayor Fiorello LaGuardia, encouraged Rickey, and public demonstrations reflected popular discontent with the prejudice that kept players such as Satchel Paige and Josh Gibson from the station their talent deserved. At last, Rickey and Chandler rallied supporters and faced down opponents, but their success didn't obscure a larger point. Baseball owners and league executives were not only timid but misguided in their understanding of the cultural role of baseball. As the 1946 vote indicates, they tended as a group to look to the past, to a time when they controlled their game without outside interference. Most of the leadership in the sport had been so inept, however, that executives such as Branch Rickey who envisioned the game's future could bring great rewards to their teams. The Dodgers' emergence in the 1950s as one of the dominant teams in the league is inextricably linked to the great advantage they had in recruiting talented black players.

In addition to his coup in signing Jackie Robinson, Rickey fortified some of the foundations established by MacPhail. The Dodger farm system grew to twenty-seven teams, and began to pay dividends after the war. Rickey's influence in the National League was especially evident during the 1946 season, when the Dodgers and Cardinals each won ninety-six games in the regular season to force the first playoff in major league history. Not only were both teams the product of Rickey's organizational principles; the Dodgers' manager, Leo Durocher, was a former member of the Gas House Gang.

In 1947, the Dodgers were forced to replace Durocher, who had been suspended by Chandler for his persistent association with gamblers. The charges had been pressed by Larry MacPhail, who was back from the war and in charge of the New York Yankees. MacPhail had perhaps never quite forgotten Durocher's undiplomatic ways during

MacPhail's time in Brooklyn; Rickey, on the other hand, thought of Durocher as a kind of prodigal son who needed latitude and understanding. Rickey may also have been mollified by Durocher's winning record. One of Durocher's greatest contributions, his last for that 1947 season, was to inform the Dodger team in no uncertain terms that Jackie Robinson had made the ballclub, and that he would play regardless of anyone's misgivings about his color.

Rickey turned to Burt Shotton to replace Durocher. The sixty-three-year old Shotton, the last National League manager to pilot a team in street clothes, arrived in Brooklyn with an undistinguished record, having guided the Phillies and Cincinnati to mediocre seasons in the 1920s and 1930s. Sparked by Robinson, Reese, Walker, and Carl Furillo, a second-year player, the Dodgers won the pennant by five games over St. Louis and faced the Yankees for the second time in the World Series. The Yankees prevailed in seven games but, unlike 1941, Dodger fans had reason to admire their team's spirit and to cherish October memories.

In the fourth game, Floyd Bevens was one out away from pitching the first no-hitter in Series history and providing the Yankees with a three-game-to-one advantage. In something less than a masterpiece, Bevens had walked in a run in the fifth inning and had put two runners on in the ninth to jeopardize his 2–1 lead. Under other circumstances Bevens would have been lifted, but Yankee manager Bucky Harris apparently thought that Bevens should be given an opportunity to attain immortality. He lost it, however, and the game, when Cookie Lavagetto pinch-hit a two-run double off the right-field wall in Ebbets Field, bringing home the tying and winning runs and knotting the Series at two games apiece. Fate was merely deferred: nine years later the Dodgers were the victims of Don Larsen's perfect game, the only no-hitter to date in Series' history.

In the sixth game, the Dodgers, facing elimination, took an 8–5 lead into the bottom of the sixth inning in Yankee Stadium. With two runners on, Joe DiMaggio drove the ball to the deepest part of left-center field. Al Gionfriddo made one of the great World Series catches, a one-hand grab on the run 415 feet from the plate, preserving the Dodgers' victory and postponing the inevitable. In the final game, Joe Page pitched brilliantly in relief to secure another world championship for the Yankees. The Dodgers could be consoled that this was their best World Series performance, and with new players arriving from the farm system, the future was bright.

In 1948, Gil Hodges, Preacher Roe, Bobby Cox, and Roy Campanella joined the cast, but the Dodgers managed only a third-place finish. The following year Duke Snider came up and provided left-handed power to a lineup dominated by right handers. Don Newcombe joined the pitching staff in 1949 and won seventeen games as well as the league's Rookie of the Year award.

The 1949 World Series brought further disappointment to the Dodgers. After splitting 1-0 wins in Yankee Stadium, the Yankees swept the three games in Ebbets Field to capture the Series in five games, beginning a remarkable string of five consecutive World Championships under their new manager, Casey Stengel.

After the 1950 season, Rickey lost control of the team to Walter O'Malley, who had been battling him in the front office for several years. Although he differed from MacPhail and O'Malley in many ways, Rickey shared with them a total commitment to running the franchise in a sound manner, which meant substantial investments in the farm system, marketing the club, and fielding quality teams.

Rickey's final years reflected the visionary and reactionary sides of his personality. He opposed night baseball and decried televised games as a drain on attendance, but he led the fight for a new major league, the Continental League, that would bring baseball to the developing urban centers of the 1960s.[17] This idea was made redundant by the willingness of the American and National leagues to expand to ten-team circuits in 1961 and 1962.

Whether the fate of the Dodgers in Brooklyn would have been different under Rickey's direction rather than O'Malley's is impossible to know. In 1948, Rickey acquired the Brooklyn Dodgers football team on the assumption that borough pride would revive a financially troubled franchise. O'Malley had opposed the purchase as an unsound business venture, and after a disastrous year Rickey was indeed forced to unload the team.[18] In 1957, in the face of the Dodgers' move to Los Angeles, Rickey, no longer with the team, charged that it would be a "crime against a community of three million people to move the Dodgers."[19] Yet sentiment alone did not dictate this shrewd businessman's judgments. He denounced the "crime," but added, "the move was [not] unlawful, since people have the right to do as they please with their property."[20] From his position outside the team, Rickey was able to extol the wonderful association between Brooklyn and the Dodgers. Had he been in a position to pursue the riches which beckoned in California, he might not have hesitated.

The Dodgers' third president after the end of the Ebbets-McKeever era was Walter F. O'Malley. As P. G. Wodehouse said of one of his characters, O'Malley "looked like a cartoon of Capital in a Labour newspaper." Less flamboyant than MacPhail and less complicated than Rickey, he was at least their equal in entrepreneurial skills. His impact on the game was summarized by Red Smith: "It had always been recognized that baseball was a business, but if you enjoyed the game you could also tell yourself that it was also a sport. . . . O'Malley was the first to say out loud that it was all business—a business that he owned and could operate as he chose. . . ."[21] For some people, that statement alone is damning, but the popular view of O'Malley is unreasonable. As Roger Kahn wrote when reflecting on his columns of the 1950s, "It amazes me to this day that once I stood in the ranks of journalists who, in the most furious words they could summon, indicted a capitalist for being motivated by a passion for greater profits."[22]

In contrast to Rickey, O'Malley was reared amid money and influence. Born in New York in 1903, he was the son of the city's commissioner of public markets. While Rickey was in Europe fighting the Great War and the Dodgers were winning their first pennant in the modern era, O'Malley was at the Culver Academy, preparing to attend the University of Pennsylvania, where he earned a degree in engineering and finished first in his class.[23] After obtaining a law degree from Fordham, he established a practice in Manhattan.

His attraction to baseball appears to have been strictly financial. Rickey thought of the sport in part as a means to instill social values in young men; O'Malley saw it as a form of entertainment. The Dodgers and Ebbets Field appealed to him because, in contrast to Yankee Stadium, their choice box seats were still available for his business clients.[24] His direct involvement with the team began when he performed legal services for the Brooklyn Trust Company, the virtual owners of the franchise after Ebbets's death.

George McLaughlin, the bank's president, appointed O'Malley the club's attorney in 1941, replacing Wendell Willkie, who had been the Republican presidential nominee in 1940, and who in 1942 was sent to Europe as Franklin D. Roosevelt's personal representative in a demonstration of American political unity. O'Malley's appointment coincided with MacPhail's departure for the army and Rickey's arrival as the Dodgers' president. In 1944, O'Malley and Rickey joined with John Smith, the president of the Charles Pfizer Chemical Company, to purchase the 25 percent of the team owned by the estate of Ed McKeever.[25] In 1944, the three purchased an additional 50 percent of the franchise, giving each 25 percent.

Smith's involvement in the club was minimal. His concern was the Pfizer company, and initially he supported Rickey completely in his efforts as the team's president. Rickey's relations with O'Malley were not so amicable. O'Malley challenged Rickey's expenditures on the farm system, the new training facilities at Vero Beach, Florida, the acquisition of the football Dodgers, and other ventures. Each man had his adherents in the organization, even down to the manager. The front office atmosphere was so poisoned by the late 1940s that Rickey believed O'Malley might have conspired with MacPhail, when he became the Yankees owner, to arrange Leo Durocher's suspension in 1947 for allegedly associating with gamblers.

As the contest wore on, O'Malley began to prevail. He rescued Rickey from a potential lawsuit after Rickey had publicly denounced two St. Louis Cardinals players and their lawyer for "avowed Communistic tendencies" because they had accepted an offer to play in the Mexican League, the first threat to the tranquility of the baseball labor market since the Federal League formed during World War I.[26] O'Malley probably shared Rickey's outrage, but didn't express it publicly and so was able to placate the players and, more importantly, their lawyer. O'Malley also prevailed in a reprise of Rickey's dispute with Cardinals' owner Sam Breadon over sponsorship of the team's radio broadcasts by a brewery. They split on the growing value of television, which O'Malley judged a new source of revenue while Rickey feared it would destroy the minor leagues.

By 1950 when Rickey's contract as president was due to expire, O'Malley had won the support of John Smith, who would die in July. For the moment, the ownership was both chaotic and partly vacant, but with O'Malley in ascendancy the stability of a new era was assured. A previous agreement by the owners provided that anyone wishing to purchase another's shares would have to match any third-party bid from an outsider. Rickey secured an offer from William Zeckendorf, a real estate magnate, of $1,050,000 for Rickey's 25-percent interest.[27] The sum was far in excess of what O'Malley had anticipated, but he matched the offer and assumed control of the franchise. As Red Smith described the transaction, "Zeckendorf received $50,000 for his trouble, Rickey got his million and O'Malley's enduring hostility. That may have been the only time O'Malley was outmaneuvered in a deal, for his financial acumen was legend. It was this talent that ultimately made him the most powerful figure in baseball, where no other quality is held in such reverence as the ability to make one and one equal three."[28]

Murray Polner's biography of Rickey only sightly overstates the nature of the clash between Rickey and O'Malley:

> In the short run, it was an argument O'Malley had to win inasmuch as they were comparing Rickey's baseball—a nineteenth and early twentieth

century slower, bucolic and pastoral sport, constant, tranquil, uninterrupted, a sentimental mirror of a world now gone—and O'Malley's vision of change and technology, of jet travel, of the surge in population and hedonism, of amoral shifts of franchises lured by more and more revenue, and other voracious appetites of television advertising.

To Rickey, baseball remained a civil religion which acted out public functions organized religion was unable to perform; O'Malley's faith rested on balance sheets and dividends.[29]

Not that Rickey was a dewy romantic wedded to a bygone era. To a considerable extent, his conflicts with O'Malley typified those between hard-nosed businessmen disagreeing over crucial investments. O'Malley, for instance, didn't interpret Rickey's signing of Jackie Robinson as a moral statement. As O'Malley told Roger Kahn, "Rickey's Brooklyn contract called for salary plus a percentage of the take, and during World War II the take fell off. It was then Rickey mentioned signing a Negro. He had a *fiscal* interest."[30]

In their differences over the farm system and Vero Beach, Rickey represented a more perceptive sense of the future than O'Malley did. Much of what O'Malley decried as wasteful spending appears to have been prudent investing which paid off for the team, especially in the 1950s. Sifting through the turmoil in the Dodgers' management of the 1940s, one finds pettiness, foresight, stubbornness, and boldness—on the part of both men. More significantly, perhaps, one finds two men ill disposed to cooperative ventures. O'Malley and Rickey each had to be in charge, and circumstances, as well as their personalities, determined the eventual winner.

The team, meanwhile, demonstrated that harmony is not essential for success. Not only did the Dodgers overcome conflicts among the club's executives, they also subordinated their own personal and sometimes racial tensions to win as they never had before. Not even changes in field managers affected them. They won pennants for Leo Durocher in 1941, for Burt Shotton in 1947, for Shotton again in 1948 after Durocher's brief return, for Charlie Dressen in 1952 and 1953, and for Walter Alston in 1955 and 1956.

After the upheaval of the 1940s, the Dodgers became a model of stability. The front office bore the stamp of O'Malley, who, by his own admission, fined club employees one dollar every time they mentioned Rickey's name. Burt Shotton was replaced as manager in part because he enjoyed a long-standing association with Rickey. And when Charlie Dressen had the temerity to ask for a two-year contract after winning 105 games in 1953, O'Malley fired him and personally chose Walter Alston, who went on to manage the club for twenty-three years. After the 1956 season, the last vestiges of the Rickey imprint were removed: Jackie Robinson was traded to the Giants.

This apparent stability was misleading. O'Malley's vision, as Polner describes it, was constrained by Ebbets Field, if not by New York itself. O'Malley had won control of one of the most valuable franchises in baseball, but feared competition from other teams. "The history of the Brooklyn club," he told Roger Kahn, "is that fiscally you're either first or bankrupt. There is no second place."[31]

In his efforts to keep the Dodgers successful, O'Malley held a problematic asset in Ebbets Field. When the stadium was constructed in 1912, its capacity of twenty-five thousand was ample, and its location was surrounded by open fields. Before long, however, its seating and location became cramped. During the 1916 World Series, the *Brooklyn Eagle* had to run the following injunction:

> Women will be requested to wear hats not more than nine feet in circumference in any part of the stand, and especially in those field boxes. This is no joke.[32]

During the dead-ball era prior to 1920, the left-field wall was over 400 feet from home plate.[33] Right field was just over 300 feet away with a 9-foot wall. The stadium was double-decked from the right-field wall around to third base. The concrete and marble structure stood impervious to the threat of fire, which had razed several ballparks constructed of wood. To accommodate growing attendance, the stadium was renovated with the addition of an upper deck of bleachers from third base around to right-centerfield. The playing field was affected in several ways. Left field was brought in to 348 feet, establishing the dimensions that favored the power-hitting team of the 1950s. The right-field wall was made more difficult to clear by the addition of a chain-link fence that raised the barrier to 20 feet.

In stark contrast to today's structures, the outfield fence was not a smooth continuous arc. Each outfield had its own fence, which intersected its neighbor at a sharp angle. Center field included a notch that extended the playing area. Right field was even more curious, as its beveled lower section met the playing field at considerably less than the customary ninety-degree angle. Line drives occasionally stuck in the chain-link suspended above the original wall or caromed bizarrely off the union of concrete and steel. The fortunate batter might send a fly ball to the base of the imposing scoreboard in the middle of the right-field wall. Such a blow, striking the "Abe Stark" sign, entitled the batter to a free suit courtesy of Mr. Stark's tailor shop. The odds against hitting the sign were sufficiently long that few suits were claimed, but the publicity helped Stark become borough president.

Until fairly recently, players honed their abilities to fit the idiosyncrasies of their home stadium. This was especially the case with Ebbets Field. Duke Snider learned to leap high for fly balls by planting his cleats in the center-field fence. Carl Furillo meticulously studied every quirk in the right-field wall. During the 1952 World Series, Casey Stengel took his young center fielder, Mickey Mantle, to the Ebbets Field outfield to recount his own experiences as a player there forty years before.

The games in Ebbets Field were distinguished also by the antics of the fans. Hilda Chester sat in the left-field bleachers and led cheers with a cowbell. The Sym-phoney taunted umpires with renditions of "Three Blind Mice." Other fans hurled fruit and vegetables at opposing players. Many teams, of course, had raucous fans, but the Dodgers' following remained unique. During World War II, as Roger Kahn points out, references to Dodger games carried over Armed Forces Radio were a staple of B movies. To root for Brooklyn at such a time became a quintessentially American activity, one of the distinguishing features of "our" side.[34]

Radio permitted the team to be followed by all the residents of the home borough. And the Dodgers had one of the finest baseball broadcasters of all time in Red Barber. Although Barber had a Southern accent (as did Mel Allen, who later broadcast Yankee games), he was quickly accepted in Brooklyn because of his ability to make fans feel present in the park. The great disciplinary threat of Roger Kahn's youth was that his parents might confiscate his Air King radio, and break his audio tie with the Dodgers.[35] The new medium provided a source of revenue through sponsors who advertised their products for a fee to the ball club. In the 1920s, owners had been concerned that free broadcasts would curtail live attendance. In fact the popularity of the teams swelled as more people were exposed to them—Brooklyn attendance, as noted earlier, attained its peak in 1947.

Radio bred other controversies. David Voigt describes the inevitable tension between live announcers and newspaper writers, the latter inveighing against the corruption of the game by overexuberant announcers who described routine or mediocre plays in exaggerated terms.[36] Even today the function of the broadcaster remains uncertain; some see his role as that of an objective reporter, while others perceive him as the team's official rooter. In fact radio descriptions of the game can serve both as a substitute to being at the park and also as an inducement to attend the game and witness the action for oneself. It is not uncommon to find people at the park listening to a superb radio announcer such as Vin Scully, whose lilting descriptions enhance the events as they unfold.

By the late 1940s, baseball had to adjust to the new medium of television, which would have a transforming effect on the game. To many the tube seemed a curiosity, a fad. Further, the technological limits of television initially made it ill-suited for baseball, unlike other sports, such as football and basketball, with their rectangular surfaces and more or less direct movement along straight lines. For such contests, television can provide a better view than a seat in the arena. Fixed cameras allow the viewer to focus on the action from an ideal angle close to the players. The contours of baseball fields and the tactics of the game itself resisted adequate television coverage. The home viewer could be afforded a detailed prospect of a part of the field or a bird's-eye vista of the entire playing area, either of which prevented him from seeing the intricate beauties of the game. The fan in the ballpark can observe a runner on first base breaking for second during the pitcher's throw to the plate, then check to see which infielder is covering second and whether the batter is attempting to hit the ball through the vacated part of the infield—this "hit and run" play is basic to the game. It can be described by a capable announcer, but the early television broadcasts provided only a partial and confused presentation of it and similarly complex plays. Nor could one camera capture the relationship between outfielders playing caroms off Ebbets Field's puzzling right-field fence and runners determining whether they should try to score.

Even weather conspired against televised coverage of baseball. Football games are held regardless of snow, rain, or freezing temperatures. Basketball is played indoors. If the weather for baseball is too inclement for spectating, it is probably also too poor to play, and the contest will be postponed. Nature can induce the football fan to stay home and watch a game on TV, but baseball is played only when the weather permits; there are seldom occasions when it profits the serious fan to stay indoors and watch from the worst seat in the house.

Another obstacle for television was the resistance to it of many owners. O'Malley was almost alone in grasping the potential for revenue. Other owners agreed with Rickey that television diminished their product by giving it free to vast numbers of casual viewers who would settle for an inferior presentation.[37] These owners clung to the notion that their teams belonged to local communities whose support determined the clubs' financial success. In fact the Dodgers and other major league clubs had been followed by fans all over America for many years. From the time baseball first gained popularity, after the Civil War, newspapers published descriptions of games and player statistics, permitting league contests and individual efforts to be followed nationwide. Television logically extended this communication by bringing the game itself into the homes of those millions of fans unable to attend major league stadiums.

Spring training and barnstorming had taken major league play into the South and the West, but these exhibitions paled in comparison to regular-season, let alone championship play, which had been confined to ten Eastern and Midwestern cities for more than fifty years. O'Malley was among the few to realize that sponsors would pay big money to advertise their products in conjunction with teams that had built a loyal following in these new markets.

One adverse effect of television has been its contribution to what Voigt describes as baseball's "Plastic Age."[38] It is often difficult to tell now, when first turning in a game, who the teams are or where they are playing, since modern playing fields are identical, perfect sectors, one-quarter of a circle imposed on a plastic surface, sometimes under a domed roof. Where, in such a game, is the place for the right-field wall in Ebbets Field, for the notch cut into center field in the Polo Grounds, for the sloping left field of Crosley Field in Cincinnati? The answer, of course, is that they have no place, and indeed no longer even exist. Fenway Park in Boston, with its high wall painted green in left field, and Chicago's Wrigley Field, with its vine-covered brick fences, remain the only truly eccentric fields.

Despite its inelegance, Ebbets Field provided a prosperous base for televising games in the mid 1950s. By 1955 the Dodgers were televising all their home games as well as more than twenty road games. That year radio and television earned the team $787,155, a sum far greater than that of every National League competitor except the Giants, who also used television generously.[39] The Milwaukee Braves, on the other hand, televised none of their games, and received just $135,000 from radio broadcasts. The disparity between the Dodgers' and the Braves' broadcast revenues helped offset the advantage Milwaukee enjoyed in gate receipts.

In 1956 the Dodgers, Giants, and Yankees gave some thought to curtailing televised broadcasts, especially of night home games.[40] When asked by the *Sporting News* if he intended to cut back on televising games, O'Malley was noncommittal and said he would examine the question after the season's end.[41] He must have liked what he saw—

the next season the Dodgers continued their policy of televising home games. Radio and television receipts increased to $888,270, while the Braves, relying on radio alone, dropped to $125,000.[42]

When O'Malley raised his objections to Ebbets Field, others countered that the Dodgers competed with themselves by giving games away on television. In the *Sporting News*, Gus Stelger of the New York *Daily Mirror* was quoted as echoing some of Branch Rickey's concerns about the impact of television: "How can't the attendance fail to fall off when the club is giving its product, the game, away?"[43]

O'Malley was rarely accused of being excessively charitable in the operation of his business, and indeed he had an idea for making television even more remunerative. Early in 1957, he began negotiating with Skiatron Corporation for pay television coverage of baseball.[44] (Early pay TV never really materialized; some plans envisioned a coin box attached to the set, on the order of some hotel and airport televisions today.) He anticipated an eventual doubling of revenues from the new technology by charging viewers one dollar, roughly the equivalent of a general admission ticket. The formula for allocating these revenues would be that the Dodgers would keep one third of the money, less 25 percent for the visiting team. When asked by Brooklyn Congressman Emanuel Celler in 1957 what such a system would generate in dollars and cents, O'Malley replied:

> We don't know, of course, what the acceptance of the public would be, but I can give you an example. One night we had 15,700 people in Ebbets Field and I asked a radio-wise man who was with me to try to get me some idea how many sets were tuned in to that particular game, and after some phone calling he came back and said, "Walter, there are about 2,400,000 sets watching this game, and you have only 15,700 people in your park." That made me convert right away.[45]

The Dodgers signed an escrow contract with Skiatron, conditioned on the reaction of other teams in the league, whose owners proved reluctant, evidently deterred by technical problems with the cable and also concerned about whether broadcasts would be regulated by the Federal Communications Commission, reducing the owners' control over the medium. O'Malley informed Celler's Antitrust Committee that no further investigation of pay television would be considered "until these hearings conclude."[46] In fact the hearings had already induced O'Malley to withdraw from the pilot agreement with Skiatron shortly before he was called to testify. Once again O'Malley demonstrated his special vision, and again his fellow owners were chary of embracing the change. Thirty years later Peter Ueberroth had to resolve pay television disputes among the owners.[47]

Television had provided the Dodgers with an opportunity to recoup some of the revenues that had fled to the suburbs along with much of Brooklyn's population. In no other way could the Dodgers have kept pace financially with the Braves while being outdrawn at the gate by over one million fans per year. The population base the Dodgers enjoyed because of television compensated for the advantage afforded Milwaukee by its municipal stadium.

There were other early signs of television's impact. On March 6, 1952, the *New York Times* carried an innocuous report from the Dodgers' spring training camp in Vero Beach, Florida:

> Norman Bel Geddes, the industrial designer who came there at the invitation of Walter F. O'Malley, president of the Dodgers, to design a 5,000 seat stadium, is designing also a new structure to replace Ebbets Field, and to be built on another site. . . .
>
> This probably is far in the future but Bel Geddes' plans are on a grandiose order. They include a retractable roof; foam rubber seats, heated in cold weather; a 7,000 car garage from which fans can proceed directly into the ballpark; automatic hot dog vending machines everywhere, including mustard; a new lighting system minus the present steel towers and a synthetic substance to replace grass on the entire field and which can be painted any color.[48]

One can hardly imagine a better stadium for the television age; later, even the pretense of green plastic "turf" would be abandoned. A drawing of this new stadium was displayed in the Dodgers' offices in Brooklyn, and was known derisively as "O'Malley's Pleasure Dome." As long as Bel Geddes's proposal seemed the only option to Ebbets Field, people were not likely to gauge how serious O'Malley was about moving his team to suitable quarters.

With the Dodgers ensconced in a weatherproof stadium in Brooklyn, they would have been in an ideal position to maintain their traditional allegiance with the borough, avoid the inevitable risks of a long-distance move, and capitalize on their national following by televising games in a way that would have poured receipts directly into the Dodgers' coffers.

For many residents and city officials in New York, the Dodgers were inseparable from Brooklyn. The team and the borough were bound through decades of history and clearly belonged to each other. But this sentiment overlooked the point that by 1950 the Dodgers and other teams belonged to communities all over the country. The Dodgers were familiar not only to those who paid admission to Ebbets field or who listened to Red Barber on radio. They were known, followed, and even loved by millions who had viewed World War II movies or later watched the team on "The Game of the Week." In the postwar era, the Dodgers had an entire country in which to move.

Indeed the whole country was moving in those days. Postwar prosperity and government policies such as the GI Bill of Rights and highway programs shifted populations out of the city. Installment-buying made cars more affordable and cars in turn permitted independence and mobility. Families which for generations had crowded into tenements now fled en masse into the suburbs.

A former Dodger official, who witnessed Walter O'Malley's revelation of the impact of the automobile on his franchise, recalled:

> O'Malley was leaving Ebbets Field after a night game. He was in his chauffeur-driven Buick. He gets stuck in a traffic jam around what is now the site of the Meadowbrook Parkway. He's going east, and he sees a tremendous amount of cars going west. "What's that?" he asks his chauffeur." "That's all the people coming home from the trots, Mr. O'Malley." Suddenly, O'Malley's ears perked. Harness racing had yet to reach its peak, its real peak. Bingo! Smack! Click! O'Malley was a shrewd man. He already saw the demographics. Hey, what were those 20,000 people doing there and not going to a ball game at Ebbets Field?[49]

In the economics of baseball before World War II, competition for the fans' money came principally from the movies. The growth of television and the automobile changed that. As fans were freed from the cities by their cars, they found new diversions, and baseball had to become more competitive.

How seriously the Dodgers attempted to make do with Ebbets Field is unclear. Parking was available for only seven hundred cars, and they, O'Malley claimed, were subject to increased vandalism.[50] His contention that the park was becoming a forbidding place was supported by Stan Lomax, a baseball writer and O'Malley's college fraternity brother:

> The scene at Ebbets Field was one of riding on the crest of a volcano. If they didn't get a new park they would have had a riot or some terrible disturbance. Especially at the midweek games—there was too darn much drinking. There were narrow aisles, the seats were too close and you had a rough, tough bunch there. If somebody threw a bottle or stabbed someone—that's all that was needed—the dynamite was there . . . with too many people in too small an area.[51]

Baseball in Ebbets Field historically included a persistent note of violence, from the physical abuse of opposing players to an ugly assault on umpire George Magarkurth. Fights among players were common and beanballs numberless.

This atmosphere was tolerable as long as baseball attracted a uniformly rowdy clientele, but the customers of the 1950s were different.

The park as a male preserve had been invaded by families with money to spend; one third of the Dodger's spectators were now women.[52] The 1950s brought prosperity and leisure, and baseball had to present itself in more genteel surroundings than in earlier times.

Ebbets Field had been a suitable ballpark when the fans composed a kind of club. Before World War II, a white, male, working-class contingent dominated the park and enjoyed one another's company as much as the game. The Sym-phoney Band, Hilda Chester, and others achieved fame in the community of rooters.

After the war, Jackie Robinson attracted thousands of blacks to the stadium, and its atmosphere changed. Peter Golenbock has recorded the recollections of fans on this point. "By the time Hilda Chester had returned around 1955, Ebbets Field was no longer the same. . . . You had a different crowd. It was no longer a unified crowd. It was more subdued, because you weren't as apt to jump up and scream across the aisle at someone because neither the white fans nor the black fans were comfortable with each other."[53] Herb Ross told Golenbock, "When the blacks started coming to the game, a lot of whites stopped coming. And the black allegiance was only to Robinson and the black ballplayers. They didn't care about the Sym-phoney or Hilda Chester or even the white players. They didn't have the history that we had. The allegiance of the blacks was not to institutions. The allegiance was to Robinson."[54]

The accuracy of such observations may be suspect; the point, however, is that many people felt ill at ease. Certainly many blacks were attending Dodger games because of their excitement about Robinson, Campanella, and other stars. In any case, while the team handled the influx of blacks with reasonable grace, the fans reacted less well. Aggravating racial tensions inside the park were changes in the neighborhood around Ebbets Field, which was located just a few blocks from the cultural heart of Brooklyn—the Botanical Gardens, Prospect Park, the Brooklyn Museum, the Brooklyn Public Library, and Grand Army Plaza, with its magnificent arch commemorating Civil War veterans.

In the early part of the century, Crow Hill, where Ebbets Field had been built, had become Crown Heights, a fashionable area. But, following the regrettable pattern of so many urban areas, this elegant community began to deteriorate when its inhabitants migrated to the suburbs. The vacuum was filled by poorer residents whose color, customs, and dialects radically altered the atmosphere of Dodger games.

With only a few hundred parking spots available, Ebbets Field now seemed an uninviting place in an increasingly unfamiliar neighborhood, and many former Brooklyn residents stopped attending games.

Radio and television enabled them to follow the Dodgers without leaving the comfort of the suburbs. At the same time, women, families, and blacks were entering Ebbets Field in record numbers, and the intimacy of the old park, which had served the prewar fraternity of fans so well, became another drawback. Stan Lomax's remarks about the potential dangers of the crowds convey the claustrophobia created by the narrow seats and aisles. Moreover, Ebbets Field's many pillars, which supported the upper decks, pressed people even closer together as fans congregated in areas with unobstructed views.

In the late 1930s, when Larry MacPhail had renovated Ebbets Field, fresh paint and new sod did the trick. As long as Brooklyn itself remained a stable community, cosmetic repairs were sufficient. By the 1950s, however, the old preserve had become obsolete. The most compelling evidence of that obsolescence is the fact that despite its small size and the perpetual drama of the pennant races of those years the park was hardly ever filled to capacity. After attendance peaked in 1947—at just over 1,800,000—the numbers dwindled, although they remained the envy of almost every other club in the major leagues. A paradoxical quality attends the Brooklyn fans of the 1950s. Since their traditional male, white, working-class contingent was augmented by women, families, and blacks, one might assume that Ebbets Field was too small to accommodate the expanded audience. But, as attendance figures attest, the number of seats was adequate. Access to the park, however, was not.

By contrast, Yankee Stadium continues to draw large crowds even though it is located in the South Bronx, an area even more likely than Flatbush to promote fears of racial tension and crime. The difference is that Yankee Stadium affords convenient parking and access roads and thus draws fans from the entire metropolitan area. It remains an urban though not a neighborhood stadium. This suggests that the racial factor has been overemphasized as a threat to the stability of the Dodgers. More important was the inconvenience of getting to the park and the discomfort inside. Ebbets Field, miles from any expressway and with limited parking and cramped seating, couldn't provide the amenities necessary for a thriving modern ballpark.

The inadequacy of the stadium ultimately concerned the larger community. The fatal constraints were not that Hilda Chester no longer fit in or that the Sym-phoney Band was an anachronism. Nor was it important that the racial composition of the fans had become more diverse. Changes in the ethnic and generational composition of a team's market have been accommodated by every durable franchise. The crucial matters are access to the park and comfort within. In those respects, Ebbets Field was beyond repair.

Some fans doubted that the Dodgers' front office was committed to retaining their allegiance to Brooklyn. Red Barber was not rehired after the 1953 season, perhaps another casualty of the battle between O'Malley and Rickey. Vin Scully provided excellent coverage, but some of his colleagues were less proficient. In his memoir of his family's involvement with New York baseball, Damon Rice recalled some frustrations from 1955:

> If it was hard following a game on TV or radio, it was equally difficult getting seats at the ball park—even though Walter O'Malley was crying the blues about lousy attendance. Whenever I tried calling up the ticket office for information, the phone would ring forever. Going down to Montague Street wasn't much better. The ticket seller would give you "the best seats left" for the game in question and move you along. It wasn't uncommon to get to the game and discover whole sections of better seats had gone begging. . . .
> On days of big games the club would often sell more general admission tickets than there were seats. People would clog the aisles looking for places that simply didn't exist, getting angrier by the minute. Many swore, "this is the last time you'll see me at Ebbets Field." . . .
> It simply didn't make any sense. When O'Malley took over the Dodgers he'd given an extraordinary amount of lip service to the notion that the fans must be kept happy at all costs. Now he appeared to be doing everything in his power to destroy public relations. I couldn't help but suspect that there might be a method to his madness.[55]

Perhaps O'Malley thought a serious upgrading of Ebbets Field would preclude the support he needed for a new stadium. Certainly, there was no way Ebbets Field could have been expanded to accommodate the Dodgers in the future. The park, constructed in a desolate section of the borough, was now surrounded by city streets that made major renovations impossible.

The changes that were straining the Dodgers in Brooklyn were proving ruinous to clubs in other cities. For fifty years, the sixteen major league teams were confined to ten Eastern and Midwestern cities. In the National League, only the Pittsburgh Pirates and the Cincinnati Reds had no competition in their own towns, and in the American League only the Washington Senators, Detroit Tigers, and Cleveland Indians enjoyed monopolies. Most cities had both an American and a National League franchise. Boston hosted the Braves and the Red Sox; Chicago, the Cubs and the White Sox; Philadelphia, the Athletics and the Phillies; St. Louis, the Browns and the Cardinals; and New York, of course, had three teams—the Dodgers, Giants, and Yankees.

The shift of population from cities to the suburbs reduced the base of financial support for most multiteam cities. Generally, a city aligned itself with one team while the other struggled to avoid bankruptcy. In the past, many teams, including the Dodgers, had come close to financial collapse, but postwar demographics introduced unprecedented challenges.

The first franchise to switch cities in response to these pressures was the Braves. In 1948 they had won the National League pennant, and drew over a million and a half fans to their decrepit park. Thereafter, they failed to finish higher than fourth place, and in 1952 came in seventh. The impact was devastating, as their total attendance plummeted to only 281,000 and the club lost over $600,000.[56]

In March 1953, less than a month before the season began, Braves owner Lou Perini petitioned to move his team to Milwaukee. The year before, Perini had blocked an attempt by Bill Veeck to move the St. Louis Browns to Milwaukee, and only a few days earlier American League owners rejected Veeck's request to move the Browns to Baltimore.[57] Surprisingly, National League owners unanimously agreed to Perini's request, and the Braves became the first team to shift cities since the Baltimore Orioles became the New York Yankees in 1903. The motion to permit the Braves' relocation was introduced by O'Malley and Horace Stoneham, the owner of the New York Giants.

The move was an immediate success. The first nine games for the Braves in Milwaukee yielded an attendance equal to the entire 1952 season in Boston. In the first year of operation in his new home, Perini realized a profit of $500,000.[58]

The first shift of an American League franchise proved more problematic. The Washington Senators had rivaled the St. Louis Browns for the distinction of being the worst franchise in the history of the league. The last thing such a team needed was competition in its local market, yet that is what the league sanctioned when it approved the Browns' transfer to Baltimore in 1953. The Browns were renamed the Orioles, and in the 1960s and 1970s became "the best damned team in baseball." The Orioles' success complicated life for the Senators, who themselves began looking for another site as early as 1956. After the World Series, Clark Griffith, the team's owner, met with city officials from Los Angeles, but kept his team in Washington until 1961 when they moved to Minneapolis. The third club to switch cities was the Philadelphia Athletics. Early in the century, the A's, led by owner-manager Connie Mack, had been one of the powers of the league. Mack, born during the Civil War, was one of the most important figures in the development of the game. Remarkably, he was still managing the A's in 1950 at the age of eighty-eight, without success. He retained ownership of the team until 1953, when, heavily in debt, he was forced to sell. The new owner, Arnold Johnson, faced the familiar problems of a decaying stadium in an unsupporting city, and within a year he moved the team to Kansas City.

The stability baseball had enjoyed for fifty years was clearly at an end, along with the oligopoly that ten cities had held for so long. The bond between a team and its community was no longer inviolate, although the borough of Brooklyn continued to believe it was exempt from these forces.

The move with the greatest impact on the Dodgers was the Braves'. The Dodgers had traditionally battled the Giants and the Cardinals, but the new competition from the Braves was seen as the beginning of a new era. In their last year in Boston, the Braves finished next to last in the league. The following year, 1953, they rose to second as young players such as Eddie Matthews, Joe Adcock, and Del Crandall provided runs for star pitchers, Warren Spahn and Lew Burdette. In 1954, slugger Henry Aaron's first year with Milwaukee, the Braves finished third, only three games behind the Dodgers. The following season the Dodgers won by thirteen games over the Braves, clinching the flag earlier than any team in league history. But in 1956 the aging Boys of Summer defeated Milwaukee's young stars by only one game.

More ominous for O'Malley than the competition for the pennant was the attendance that the Braves were generating in Milwaukee. The Dodgers continued to attract about a million fans to Ebbets Field, but they needed significant television revenues to ensure a profit. The Braves, meanwhile, playing in County Stadium in Milwaukee became the first team to draw more than two million. As O'Malley saw it, Milwaukee's record attendance would permit the Braves to sign the best young players available, and within a short time eclipse the Dodgers and the rest of the League. O'Malley's association with his team began when the club was directed by the Brooklyn Trust Company, and he well understood the tenuous nature of financial success in baseball. Later, reflecting on that time, he told Roger Kahn:

> . . . we can't even afford a few years of this. The Braves will be able to pay bigger bonuses, run more farm teams and hire the best scouting talent.[59]

His premonitions were accurate—the Dodgers fell to third place in 1957 behind the Braves, who won the World Series that year and in

1958 repeated their league championship. The Dodgers were not to prevail again until 1959, in a decisive playoff game held in the Los Angeles Memorial Coliseum, where the Dodgers drew up to ninety thousand for a single game and easily surpassed the Braves' attendance and revenues.

The only chance Brooklyn had to keep the Dodgers was to construct a variation of Bel Geddes's proposed stadium at the junction of Atlantic and Flatbush avenues, the terminus of the Long Island Rail Road and also the meeting point for two subway lines.

O'Malley was determined to build the new stadium with private capital, the first stadium so financed since Yankee Stadium was erected in 1923. The Braves and Orioles had moved into public stadiums, and the A's played in an old minor league park in Kansas City. What O'Malley needed from New York city officials was not money for the stadium but access to the site. Through urban renewal provisions specified by Title I of the 1949 Federal Housing Act, businesses presently operating at Atlantic and Flatbush avenues could have been relocated and the land made available to the Dodgers. City administrations all over the country had used public funds to attract teams and the revenues and status they generated. The challenge for New York officials was to exercise their power hold on to one of the most attractive franchises in the game.

3

Robert Moses and the Atlantic-Flatbush Proposal

The village of Brooklyn at the westernmost end of Long Island was incorporated as an independent city in 1834, five years before Alexander Cartwright's mythical first game was played at Cooperstown. Brooklyn remained autonomous until after the Civil War, when the visions of two people changed its status.

One was John Augustus Roebling, a master of mathematics, engineering, music, and other pursuits.[1] Roebling proposed the construction of a bridge across the East River, linking Brooklyn and Manhattan and freeing passengers and commerce from their reliance on the ferry, a problematic mode of transportation, especially during the winter. Roebling was surveying a site on July 6, 1869, when a ferry crushed his foot. Complications set in, and several weeks later he died of blood poisoning. His son, Washington Roebling, took his place, and construction on the bridge began on January 2, 1870. The younger Roebling was himself seriously injured two years later when his underwater inspection of a caisson afflicted him with the bends. Thereafter he was confined to his home, where he oversaw the construction from a window through fieldglasses while his wife conveyed his instructions to the crew laboring on the bridge.

Despite numerous delays, work progressed, and in 1883 the project was completed. The Brooklyn Bridge was the longest suspension bridge in the world, and with the possible exception of the Suez Canal, which had opened in 1869, it was the greatest engineering feat of the age. As John Roebling had promised, the bridge indeed strengthened the ties between Brooklyn and Manhattan and also symbolized the vision of the other key figure in Brooklyn's fate.

Andrew Green was a civic leader who foresaw an imperial city, a

Greater New York.[2] The census of 1890 established New York (that is, Manhattan and parts of the Bronx) as the largest city in the United States, with a population of one and a half million residents.[3] Philadelphia and Chicago followed, each with populations in excess of a million, and then Brooklyn, with over eight hundred thousand. The merging of Brooklyn and Manhattan was inspired not only by dreams of grandeur, but also by pragmatic concerns, especially financial ones. Although a few metropolitan services were shared by Brooklyn and Manhattan, the cities duplicated other services that more efficiently and economically could have been shared.[4] Green and others lobbied for consolidation of Brooklyn and Manhattan, but some residents and officials in both cities were skeptical. Brooklynites deplored Manhattan's sophistication, while to New Yorkers Brooklyn's slower pace and bucolic traditions made it seem the home of rural bumpkins.[5]

Nonetheless, by 1897 the forces of consolidation triumphed, and on May 4 the state legislature established a new charter for the city of New York incorporating Manhattan, Brooklyn, Queens, the Bronx, and Staten Island as boroughs or political subdivisions of a new City of New York. The new charter increased the area of the city three times and doubled its population, but it couldn't erase long-standing conflicts between the boroughs.[6] Thus, even though the charter seemed to bestow significant power on the mayor, political influence was shared among many competing factions within the city. And the memory of William Marcy "Boss" Tweed, who plundered New York during the 1870s, was too fresh to permit much concentration of power.[7]

A new alignment of political interests was codified in a revised charter passed by the state government in 1901. Its principal reform was a reduction of the mayor's power by the creation of the Board of Estimate comprising the mayor, the president of the city council, the comptroller, and the five borough presidents.[8] In order to balance borough and city interests, each borough president was given one vote while the citywide officials had two.

The Board of Estimate was an attempt to meet one of the great civic challenges, classically formulated by James Madison: "In framing a government which is to be administered by men over men, the great difficulty lies in this: you must first enable the government to control the governed; and in the next place you must oblige it to control itself."[9] The Madisonian balance prescribes a golden mean in which government has sufficient power to act in the public interest but not so much power that those interests can be abused. The 1901 reform charter represented the interests of each borough directly in the Board of Estimate and protected local communities against a mayor supported by powerful political forces such as Tammany Hall. At the same time, when decisive action was required in a single borough (as in approving land for a new stadium in Brooklyn), the federated organization of city government loomed as a serious obstacle.

When power is diffused in this fashion, it is up to the skill and willingness of key public officials to bring diverse groups to a point of accommodation. To realize Walter O'Malley's request for land for a new stadium, the support of two New York officials was indispensable: Robert Wagner and Robert Moses.

Wagner's father, Robert Sr., had been a United States senator who, though a liberal reformer, was an important member of Tammany Hall.[10] The future mayor attended an elite prep school, then received his bachelor's degree from Yale. Several years later, he returned to Yale Law School, graduating in 1937. A few months later, Wagner was elected to the state assembly, with his father's help. He was decorated for service in Europe during World War II, and returned to New York, serving—with Tammany's backing—on several city commissions and then as Borough President of Manhattan. In 1953 he was elected mayor of New York, and with consummate political skill walked the line between obeisance to Tammany and his commitment to reform. He strengthened the office of mayor by guiding a new city charter to realization. He was reelected in 1957, despite the opposition of the disaffected Tammany organization, irked by his support of reforms, and of the Democratic organizations in the other boroughs.

No less formidable than Wagner was Robert Moses, perhaps the most powerful figure in twentieth-century New York City government.[11] Moses's stature was the result not of electoral strength—he was defeated in his only campaign, when he ran for governor in 1934—but of his administrative brilliance and of his ambition, which seemed limitless. Moses transformed New York through his control over its parks, highways, and urban renewal, all directed by autonomous government agencies which he headed. His image was of the tireless public servant committed to providing recreation for the working classes, devising safe and efficient routes of commerce for business, and eradicating slums that trapped the poor. As he pursued those goals, Moses antagonized powerful public officials, including Franklin D. Roosevelt, Fiorello LaGuardia, and Robert Wagner, none of whom dared take his measure for fear of public outrage.

When Walter O'Malley sought Moses's support for land on which to build a new stadium, he cited Title I of the Federal Housing Act of 1949, a major piece of social engineering secured by the Truman Administration through the Senate leadership of Wagner—the mayor's fa-

ther—and Robert Taft of Ohio.[12] The act was intended in part to eliminate urban slums by providing a local agency with federal funds to
purchase property in desolate areas and either construct a public project or else sell the land to a private developer whose construction would
conform to a larger "public purpose," a phrase that eventually became
crucial for the Dodgers. Since Moses was chairman of the mayor's
Committee on Slum Clearance, he was the chief administrator of the
city's Title I authority and the ultimate interpreter of "public purpose."

On August 10, 1955, O'Malley sent a letter to Moses summarizing
his wishes for specific plots of land at the Atlantic-Flatbush site. On
August 15, Moses replied sharply that "I can only repeat what we have
told you verbally and in writing, namely, that a new ball field for the
Dodgers cannot be dressed up as a Title I project." He went on to
rebuke O'Malley:

> I don't see how you can have the nerve to indicate that you have not
> received proper support from public officials involved. Every reasonable,
> practical and legal alternative we have suggested has been unsatisfactory
> to you. The record shows we have made many suggestions, even though
> it is not part of our official duties. Let's be honest about this. Every con-
> ference we have attended over several years including the last one at the
> News began with a new Dodger Ball Field as the main, primary objective
> with other improvements a peripheral and incidental purpose. . . .
> However, if the Board of Estimate on the advice of the Borough Presi-
> dent of Brooklyn wants to put through a reasonably sensible plan for high-
> way, railroad terminal, traffic, street, market and related conventional public
> improvements, and incidentally wants to provide a new Dodgers Field at
> Flatbush and Atlantic Avenues, you can be sure that my boys will fully
> respect the wishes of the Board and do everything possible to help.[13]

The letter suggests that Moses concurred with O'Malley about the
appropriateness of waiting for a recommendation from the Board of
Estimate. The Board subsequently authorized a survey of the site
bounded by DeKalb Avenue, Vanderbilt Avenue, Sterling Place, and
Bond Street. In the wake of that action, Moses wrote a candid letter to
Borough President, John Cashmore, who sat on the Board, indicating
his personal assessment of O'Malley's request.

"We have no confidence," Moses wrote, "in Walter O'Malley's scheme
to put a Dodger Field at the Brooklyn terminal of the Long Island Railroad," He added, "We see no prospect of an alternate Dodger location
in Brooklyn. We think the Dodgers might be an incident in the acquisition of the Jamaica Track and its development, largely for housing for
several income groups."[14] Moses's letter suggests that the real concern
of New York officials was how to keep the Dodgers in the vicinity,
much as the football Giants and Jets retain a nominal affiliation with

New York, even though they now play in Giants Stadium, in Ruther-
ford, New Jersey, across the Hudson River from Manhattan.

The public authorities Moses headed, such as the Triborough Commission, raised their own money by selling bonds to the public and
thus were insulated from the controls that legislative bodies customarily exercise through the budgetary process. Since Moses's tenure was
established by law, he was protected from removal by the mayor or
governor for policy differences. Finally, Moses himself had drafted the
laws that established these public authorities and could be confident
that no court would find his conduct illegal. While even a mayor with
the power of Robert Wagner was constrained by the city council, the
Board of Estimate, the courts, and state and federal governments, Moses
was virtually exempt from the conventional limits of power.

After a meeting with O'Malley and Mayor Wagner, in which the
Dodgers' intentions to replace Ebbets Field were discussed, Moses reiterated his objection to O'Malley's plan to relocate at Atlantic and Flatbush, stressing the legal constraints on land use under urban renewal
provisions of the Federal Housing Act. When O'Malley emphasized his
interest in the site, Moses replied, "Then, if you don't get this particular site you'll pick up your marbles and leave town?"[15] O'Malley responded that he had no such intention, but later he speculated at a
news conference that New York might become a one-team town, with
the Giants' fate inevitably determined by the Dodgers'.

Moses's emphasis on the legal constraints of Title I may have been
somewhat disingenuous. In his biography of Moses, Robert Caro describes the significance of the housing law:

> Title I of the Housing Act of 1949 extended the power of eminent do-
> main, traditionally used in America only for government-built projects, so
> drastically that governments could now condemn land and turn it over to
> individuals—for them to build on it projects agreeable to the government.
> Under Title I, whole sections of cities could be condemned, their residents
> evicted, the buildings in which those residents had lived demolished—and
> the land turned over to private individuals. Here was power new in the
> annals of democracy. And in New York, that power would be exercised
> by Robert Moses. "In my opinion," urban expert Charles Abrams was to
> say, "under present redevelopment laws, Macy's could condemn Gim-
> bels—if Robert Moses gave the word. . . ."[16]

In Robert Moses, O'Malley had met a man driven by a grand vision
for the future, one that in a sense expanded the visions of Green and
Roebling. Roebling had constructed a single bridge that tied Brooklyn
to Manhattan and made consolidation inevitable. Moses not only constructed bridges and tunnels that linked the boroughs more closely; he

also designed the parks and roadways that lured the city's populace to surrounding areas. He was almost solely responsible for opening Long Island to the immigrant and working-class residents of New York's inner-city tenements. Before Moses built a network of expressways, Long Island had been a preserve for a few wealthy families. With only a few roads used almost exclusively by the rich to reach their estates and clubs, the enormous expanse of land had no adequate system of transportation for masses of travelers who longed to visit beaches and parks as an escape from the pressures of the city.

By the end of World War II, however, New Yorkers were using Long Island not only as a vacation spot, but were moving there to enjoy the comforts of suburban life. Moses's concept of a greater New York included counties whose residents commuted many miles each day to and from their homes to jobs in the city. Few of these people found it appealing to joust for one of the seven hundred parking spaces at Ebbets Field, especially since fans who no longer lived in Brooklyn could follow the team on radio or television.

To Moses, the Dodgers were a small element in his much grander scheme to restore New York. Moses effected vast changes in New York by relying, to a great extent, on the Title I authority which he wielded. In case after case, as Caro demonstrated, where the law impeded Moses, Moses found a way around it. Laws were rewritten, superseded by special legislation, or simply ignored. When Moses told Walter O'Malley that Title I of the Federal Housing Act would not permit the use of land for the construction of a baseball stadium, what he meant was that he did not wish the land used in that way.

Moses's antipathy to the Dodgers' proposal was evident enough, but his reasons remain somewhat elusive. One factor was that Brooklyn itself, a single borough in a vast, expanding metropolis, didn't count for much. Another reason was that Moses's idea of recreation for city residents involved more participatory activities than watching others play baseball. Jones Beach and Riverside Park were both developed by him. He wanted New Yorkers to have access to places where they could swim or picnic. Facilitating mass spectatorship was not part of his plan.

There was yet another possible reason for Moses's opposition to O'Malley. The renovation of the Atlantic-Flatbush site would have served to improve the Long Island Rail Road terminal. As Borough President John Cashmore remarked, "The Long Island Railroad has purchased millions of dollars worth of new passenger cars, none of which can come into the Atlantic Avenue depot because of track curvatures and platform limitations. This is one problem that we would like to solve so that our people and merchants can have the benefit of modern terminal facilities and the use of the finest rolling stock on the line." [17]

If Moses was ambivalent about the Dodgers, he was in open competition with the LIRR. Passenger trains provided mass transportation, precluding the need for the network of roads to which Moses was committed. Expressways for commercial traffic and roadways for passenger cars were a major factor in Moses's power, giving him control over thousands of jobs and making him the dispenser of large contracts. In a letter of August 26, 1955, Moses wrote to John Cashmore that he had "complete confidence in the rehabilitation of the terminal area." [18] But this expression of support conflicts with many previous decisions that frustrated the development of mass transit for the city and its surrounding communities.

Moses's power was perhaps at its pinnacle in 1955. He ushered in the year by announcing an alliance with a former adversary, the Port Authority, which issued with him a "Joint Study of Arterial Facilities." The plan included the construction of three bridges and connecting roadways, in effect sealing the fate of transportation between New York and its suburbs by favoring the automobile and making money unavailable to the railroads, which desperately needed modernization.

Robert Caro writes that in January 1955 the Triborough and Port authorities had a combined fund-raising capacity of $1.25 billion. This money financed a strategy that precluded the upgrading of rail and subway transportation. Commuters left the trains and joined the massive traffic jams which no construction effort could prevent. Later in 1955, construction on the Long Island Expressway began with no provision for either train or bus travel, thus further restricting Long Island commuters to the automobile. At a time when American economic power was unrivaled and the automobile king, Moses decreed that Long Island commuters would henceforth depend on the automobile, not only in the 1950s but also in the foreseeable future.

The implications of this decision were devastating for the Dodgers. Their fans who had left Brooklyn for Long Island, New Jersey, and suburban New York required swift and convenient means of transportation to see the Dodgers play. Moses, however, condemned them to the car, a headache not only because drivers must contend with the hazards of traffic and road conditions but also because of Moses's tollbooths, which created traffic jams at the bridges and tunnels, turning an evening's drive into a nightmare.

Had Moses supported the renovation of the LIRR terminal, his more general transportation policy would probably have militated against a new baseball stadium in the borough. Unlike many other cities, especially in the West, New York isn't easily traversed by a lattice of roadways. The city is composed of islands—the Bronx is the only borough that is a part of the contiguous United States. Fans from New Jersey would have had to cross the Hudson and East rivers to reach Brooklyn or take a route through Staten Island and across Moses's greatest accomplishment, the Verrazano Narrows Bridge. Even from Queens and other nearby places on Long Island, the choice was to fight the traffic rush to get home, only to turn around and head back to the city, or else wait in town, attend the game, and then face a long drive home. Well before gasoline shortages and higher energy prices, the utility of the automobile in New York was limited.

In conjunction with Moses's shaping of transportation patterns, New York in the 1950s was undergoing dramatic demographic changes. As one of the oldest and most settled parts of the city, Brooklyn afforded few opportunities for families to build new homes under incentives such as the GI Bill, which was helping spur development in new communities all over America. Commercial and residential development had long ago consumed available land in Brooklyn and thus excluded the borough from many postwar building opportunities.

As Brooklyn's wealthier white population joined the migration to suburbia, newly arriving black and Puerto Rican residents filled the vacuum in the older urban neighborhoods. From 1950 to 1960 net migration from Brooklyn resulted in a decline of 476,094 whites and an increase of 93,091 non-whites.[19] During that decade, Brooklyn's population changed as follows:

	Total	White	Non-White
1960	2,627,391	2,245,859	381,460
1950	2,738,175	2,525,118	213,057

The figures indicate that although in 1960 Brooklyn was still an overwhelmingly white borough, its racial composition was changing, and in specific communities the changes were profound. Some of the residents interviewed by Peter Golenbock claim that the demographic shifts of the 1950s permanently altered their perception of the borough.

According to Golenbock, another cause of disenchantment with Brooklyn was postwar prosperity: "When these children grew up, Brooklyn—with its concrete barrenness in winter and heat brutal enough to fry eggs in summer—was no longer acceptable. The sights of these new Americans were set on something better: the verdant, quiet suburbs of Long Island, Staten Island, and New Jersey."[20] Traditional practices, customs, and social values were also evolving. What had been familiar and reliable was now alien and threatening. Institutions, such as the *Eagle*, the newspaper that had sustained Brooklyn's identity, were passing from existence, and decisions made by government, industry, and finance drove the point home.

Could O'Malley have adjusted to those forces and used his impressive business skills to keep the Dodgers thriving in Brooklyn, as the Chicago White Sox, Detroit Tigers, and Boston Red Sox remained in traditional settings despite similar obstacles? To answer that question, one must place the actions of the Dodgers' president in the specific context of the urban reorganization that so affected New York in the 1950s.

The Dodgers were a commercial enterprise whose market was in the midst of upheaval, and the record suggests that O'Malley was more a reactor to events around him than an instigator. The passionate anger Walter O'Malley continues to inspire in old Brooklyn Dodger fans stems perhaps from the fact that the team's departure to Los Angeles demonstrated more conclusively than any other event that Brooklyn had been irrevocably transformed. Roger Kahn has written:

> After World War II, Brooklyn, like most urban settlements, began a struggle to survive. Brooklyn had been a heterogeneous, dominantly middle-class community, with remarkable schools, good libraries and not only major league baseball, but extensive concert series, second-run movie houses, expensive neighborhoods and a lovely rolling stretch of acreage called Prospect Park. For all the outsiders' jokes, middle-brow Brooklyn was reasonably sure of its cosmic place, and safe.
>
> Then, with postwar prosperity came new highways and the conqueror automobile. . . . For $300 down one could buy a Ford, a Studebaker or a Kaiser, after which one could drive anywhere. . . . Whole families left their blocks for good. They had been overwhelmed by the appeal of a split-level house (nothing down to qualified Vets) on a treeless sixty-by-ninety foot corner of an old Long Island potato farm. . . .
>
> Exodus worked on the ethnic patterns and economic structure and so at the very nature of Brooklyn. As old families, mostly white, moved out, new groups, many black and Puerto Rican, moved in. The flux terrified people on both sides. Could Brooklyn continue as a suitable place for the middle class to live?[21]

Statistics indicate that Brooklyn remained predominantly white and middle-class through the 1950s and 1960s, yet statistics did not dispel the impressions of lifelong residents that their world was passing. On August 16, 1955, Walter O'Malley revealed that a further change would occur during the 1956 season. He announced that the Dodgers would

play seven of their home games in Roosevelt Stadium in Jersey City,
New Jersey. O'Malley had discussed for some time the need for a new
stadium, but this was the first indication that the team's exclusive association with Brooklyn was not inviolate. Since the Jersey City facility
had a capacity of only twenty-five thousand permanent seats, significantly fewer than the thirty-two thousand seats in the venerable stadium in Brooklyn, it was not a serious contender to replace Ebbets Field.
O'Malley's purpose in making the announcement appears in part to
have been to create a dramatic effect. As Arthur Daley, the *New York
Times* sports columnist, put it, "He [O'Malley] was just giving a deft
jab of the needle to the Brooklyn citizenry." Daley then added his own
opinion: "O'Malley's frantic reach for the extra dollar had gone beyond
the bounds of decorum and common sense."[22] O'Malley proposed that
the Dodgers build a new stadium with their own money, in contrast to
the trend developing in the 1950s to finance stadiums publicly. The
Dodger president indicated that the Dodgers would purchase the land
for the stadium site although as O'Malley stated, "We do need help
from the city to acquire the necessary land at a reasonable price."[23]

"The necessary land" was part of a proposed redevelopment project
in the vicinity of Atlantic and Flatbush avenues, a little over a mile
from Ebbets Field. The land was occupied by the Long Island Rail Road
Depot, the Fort Greene Market, and a number of small businesses.
John Cashmore anticipated that the market could be moved and a new
depot created for the railroad, with sufficient land left to construct a
new stadium for the Dodgers with parking for two thousand cars and
the advantage of direct access to the LIRR as well as to two major
subway lines.

Horace Stoneham, owner of the Giants, sent a telegram indicating
that his team also needed help with traffic and parking problems.[24]
Stoneham gave no indication at the time that the Giants might also
have to consider leaving New York, although he did support O'Malley's efforts for a new stadium. O'Malley himself brought the Giants
into the picture by claiming, "This problem is bigger than the Dodgers
alone. It's unlikely that one club or the other would move. You'll find
that the two will move. If one team goes, the other will go." Robert
Moses repeated his objections to using Title I authority for purposes of
a new stadium at Atlantic and Flatbush, but he did suggest that those
powers might be employed somewhere else in New York. O'Malley
retorted: "Other cities have gone further. They've built the ball parks
and rented them to clubs. Other cities can't understand why we can't
do it when they can. We have to start trying to find out how it can be
done, not why it can't be done."[25]

The Jersey City gambit is best understood as a ploy to force New York City officials to comply with O'Malley's clear preference to stay in Brooklyn. Years of pointing to the deficiencies of Ebbets Field had resulted in some meetings, but no action appeared in the offing. Scheduling games in Roosevelt Stadium focused the attention of the community on the Dodgers' dissatisfaction with their decrepit park. The Jersey City announcement itself came within days of Robert Moses's blunt rejection of the Dodgers' proposal to use Title I authority to make land available at Atlantic and Flatbush. Many accounts of the incidents that followed claim O'Malley tried and failed to commit city officials to build a new stadium for the Dodgers with public funds, but in fact O'Malley repeatedly insisted on using his own capital to build the new facility. The subsequent attempt by the city government to keep the Dodgers in Brooklyn centered on developing a multipurpose sports center operated by a municipal agency with the Dodgers as one of several tenants. This idea clashed with O'Malley's determination that as a private enterprise the Dodgers must be in control of their own facility.

A preliminary step to evaluate O'Malley's plan was proposed by John Cashmore at the August 19 meeting. He suggested that $50,000 be appropriated by the Board of Estimate for an engineering study of the Atlantic-Flatbush site. Mayor Wagner agreed to support the proposal at the next meeting of the Board, scheduled for the following week. Within four days Cashmore's request had doubled to $100,000 for a study of 110 city blocks of roughly 480 acres.[26] A new stadium for the Dodgers was only one item proposed to the Board of Estimate. Also mentioned were the need for commercial and residential redevelopment of substandard and unsanitary dwellings, the problem of traffic congestion, a possible new terminal for the Long Island Rail Road, and relocation of the Fort Greene Meat Market.

Abraham Beame, then Director of the Budget for New York City, cited these objectives in a report to the Board, and added:

> The Brooklyn National League Baseball Club has expressed a desire to relocate in this area and much public discussion has centered around this fact. It is expected that the consultant will fully explore the possibility of locating a stadium in this area. Representatives of the office of the Borough President of Brooklyn have given assurance that if this should materialize, the Brooklyn National League Baseball Club will finance all costs including site acquisition and construction costs.[27]

Elsewhere, Beame noted that "The assessed valuation of the property on which the proposed stadium was shown is $6,386,900."[28] That figure is important, since in his letter to O'Malley of August 15, 1955,

Moses had asked: "If you need only three and a half acres of land, if it is indeed distressed property, if you have a million dollars in the bank, if you have railroad easements, if you really want to stay in Brooklyn, why don't you buy the property at a private sale?"[29] In his book, *Bums*, Peter Golenbock makes a similar point, emphasizing the historic relationship between the Dodgers and the Brooklyn Trust Company and reasoning that any funds needed to secure land for a new stadium would have been readily forthcoming from that bank.[30] The problem with a private purchase was that, while there was only one buyer, there were many sellers. The Dodgers needed the city or state to arrange the land into a single package with a negotiable price. Given the purchaser's fame and also his intentions, the value of individual plots of land would quite likely have risen spectacularly.

In a simpler time, Ebbets had purchased the land for his stadium in precisely the fashion recommended to O'Malley. Golenbock describes the problems he encountered:

> When he checked the deeds to see who owned the land, Ebbets discovered forty claims of ownership, either by deed or squatters rights. He formed a corporation and, disguising his true purpose, bought the first parcel in 1908. Midway through the three years it took him to secure the other parcels, word leaked out as to his objective, and several of the plot owners hiked their prices sharply. By the end of 1911, he had been able to acquire the entire area except for one parcel—he had been unable to locate the owner. Private dicks traced the man first to California, then to Berlin, then to Paris. Ultimately, he was found—in Montclair, New Jersey.[31]

Golenbock reports that the owner of the critical plot was unaware of Ebbets's purpose, and was greatly amused that someone wished to purchase land in Pigtown. He was happy to sell his plot for $500. The chances that in the 1950s Walter O'Malley could have formed a secret corporation and purchased land for a new stadium without the media and then the public finding out are too remote to be entertained seriously. Such a scenario would have brought a real estate boom to the Atlantic-Flatbush area.

In testimony before the Antitrust Subcommittee of the House of Representatives in 1957, Walter O'Malley reviewed the history of his efforts to construct a new stadium in Brooklyn: "We studied altogether five different sites, but each time we came up with the proposition that a ball club could not acquire the land because the land would have to be condemned under an eminent domain law. Mr. Moses met with us on a number of occasions and guided our thinking."[32] The ubiquitous Mr. Moses correctly pointed out that the land could not be assembled without government invocation of eminent domain—the authority to

take private land for a public purpose—and Moses himself was opposed to invoking that power in this case.

O'Malley also faced opposition from the chairman of the House Judiciary Committee, Emanuel Celler, who also chaired the Antitrust
Subcommittee and happened to represent Brooklyn. After O'Malley had
outlined the benefits to accrue from a complete renovation of the
Atlantic-Flatbush site—a new railroad terminal; a new meat market;
elimination of traffic congestion; and a new stadium for the Dodgers—
Celler focused on the final point: "Do you think that a baseball club
which has made the profits that your club has should be benefited by
acquisition of land by eminent domain?"[33]

O'Malley replied that the ball club was not benefiting from eminent
domain but would simply be acquiring land already gathered through
eminent domain. O'Malley probably was making a distinction without
a difference, though earlier in his testimony he argued that the
Atlantic-Flatbush site would be renovated as part of a large public purpose—the various improvements being considered along with the new
stadium.

The evidence indicates that O'Malley was prepared to purchase land
and construct a new stadium for the primary if not sole purpose of
keeping baseball in the borough of Brooklyn. To that end, he needed
the cooperation of the city government—not to build him a stadium
but to condemn private land, compensate the original owners, and then
sell the property to the Dodgers. The private market alternative was
barely feasible in 1910; in the 1950s it was impossible.

4

An Elusive Championship

The final years of the Brooklyn Dodgers represented a supreme irony. Between the foul lines the team achieved a level of success equaled only by the best clubs in the game's history. But in other contests of greater significance the Dodgers were repeatedly stymied.

While Walter O'Malley's announcement that the Dodgers would play some home games in New Jersey came as a shock, the team's success in 1955 was also somewhat surprising. The previous campaign, the first year of Walter Alston's tenure, had been disappointing. Alston had replaced Charlie Dressen, who in three years as the Dodgers' manager had come a run shy of winning three pennants.

Dressen's replacement of Shotton after the 1950 season represented one of the first changes of the O'Malley era. Dressen had managed Cincinnati in the 1930s with limited success, and had coached for the Dodgers during the 1940s. Like Leo Durocher, Dressen had a fondness for gambling that led to his brief firing from the Dodgers by Branch Rickey in 1943. After the war, MacPhail lured him from the Dodgers to coach the Yankees.

In his rookie year as manager, Dressen piloted the finest talent in Brooklyn's history. Campanella, Robinson, Reese, and Snider all eventually entered the Hall of Fame. Some argue that first baseman Gil Hodges belongs with them, and outfielder Carl Furillo is perhaps just a notch behind. A weak left field was shored up during the season when the Dodgers acquired Andy Pafko from the Chicago Cubs.

The pitching staff included Newcombe, who won twenty games, and Roe, who won twenty-two. They were joined by Ralph Branca and Carl Erskine. The only flaw was a lack of depth, which proved fatal during the stretch drive.

On August 11, the Dodgers enjoyed a deceptive thirteen-and-a-half-game advantage over the Giants, who were led by rookie Willie Mays, Monte Irvin, Alvin Dark, and Eddie Stanky. Sal Maglie and Larry Jansen each won twenty-three games that season, and Durocher, now managing the Giants, used his bench masterfully. The Dodgers may have been better than the Giants, but not by thirteen games, and in the middle of August, the Giants went on a tear, steadily cutting into the Dodgers' lead. With two games left in the season, the teams were tied, and the Dodgers had to rally to win the final game of the season against Philadelphia, aided by a brilliant catch by Jackie Robinson.

The three-game playoff focused the attention of all of baseball on New York. The winner would play the Yankees for the world championship. The Giants won the opener 3–1 in Ebbets Field, but the Dodgers rebounded with a 10–0 drubbing in the Polo Grounds. The decisive game was played in the Giants' park on October 3. For this most dramatic game, the Polo Grounds was only about two-thirds full. Weather was perhaps a factor, since the skies were so ominous that the lights came on in the third inning. The Dodgers scored a run in the first inning and their weary ace, Newcombe, made it stand until the bottom of the seventh. The Dodgers scored three runs off the Giants' best hurler, Maglie, in the top of the eighth, and Newcombe retired the Giants in order in the bottom of the frame to maintain Brooklyn's 4–1 lead.

In the ninth, the Dodgers went out in order, and Newcombe took the mound three outs from the pennant. Shortstop Al Dark led off with a single to right, and outfielder Don Mueller followed with another single through the hole. First-baseman Monte Irvin, a power hitter, fouled out to Hodges, bringing outfielder Whitey Lockman to the plate. Lockman doubled home Dark and put the tying run on second. Dressen pulled Newcombe and called in Ralph Branca to quell the rally. Branca had given up a home run to third-baseman Bobby Thomson in the first game, but Dressen decided to pitch to him rather than put the winning run on base intentionally with Mays on deck. Branca threw a called strike and, in order to set up a sequence of pitches, aimed the second pitch inside—but not far enough. Thomson pulled the most famous home run in the history of baseball into the first few rows of the left-field seats.

For Brooklyn fans the images of that moment are still catastrophic: Pafko pressed against the left-field fence watching helplessly as fans fight for the ball; Branca making the long walk back to the clubhouse behind center field; other players walking aimlessly in disbelief. The dramatic reversal joined Mickey Owens's passed ball in the 1941 World

Series in the lore of bitter disappointments that had become part of the Dodgers' mystique. A team without character could not have rebounded from the 1951 season as the Dodgers did. Next year they won the pennant by four games over the Giants, refusing to fold as the fierce tempers of Durocher and Dressen enlivened the interborough rivalry.

Joe Black, the Dodgers' Rookie of the Year, posted a team-leading 15–4 record with an e.r.a. of 2.15. He pitched a complete-game victory over the Yankees in the opener of the World Series. The Yankees took the second game in Ebbets Field, then dropped two of three in Yankee Stadium. The teams returned to Brooklyn for the final two games, the Dodgers needing just one win for their first world championship. The two pitchers, the Yanks' Vic Raschi and the Dodgers' Billy Loes, pitched shutouts until the Dodgers scored a run in the bottom of the sixth. The Yankees rallied for two in the seventh, and a home run by Mickey Mantle in the eighth secured a 3–2 win for New York. In the seventh game, Joe Black pitched well, but the Dodgers managed only two runs and New York clinched its fourth straight World Series with a 4–2 win. Again a season had ended in disappointment, even though the Dodgers had pressed one of the great teams of all time to the limit.

In 1953, Brooklyn established its dominance over the league. Dressen's team won 105 games that year, a club record. Carl Furillo took the batting title, while Campanella hit 41 homers and drove in 142 runs to top the league. Carl Erskine, with 20 wins, was the only pitcher with more than 15 on a staff that was balanced and deep. But the World Series was another let down. The Dodgers won the third and fourth games to knot the Series at two, but the Yankee machine pounded out an 11–7 win in the fifth game before returning to Yankee Stadium. With two runs in the ninth, the Dodgers rallied to tie the sixth game, but Billy Martin drove in the winning run in the bottom of the ninth to give New York its record fifth straight championship.

Dressen's three years at the Dodgers' helm were as successful as any equivalent period in the team's history. The only misfortune was that they coincided with a slightly better team in the Bronx. This period in the 1950s sealed Brooklyn's attachment to the Dodgers, a team too good to merit despair but not quite able to fulfill the aspirations of its followers. Had Tantalus received his punishment in the 1950s, he would have been sent to Brooklyn. Dressen was popular with his players, and, though hardly a diplomat, he successfully guided the team through the difficult phase when black players less heralded than Jackie Robinson replaced several established Brooklyn starters.[1] Dressen's remarkable record was outweighed by his impolitic request for more job security than was provided by the one-year contract that was one of the firmest traditions of the O'Malley era, and he was released.

Walter Alston, after a major league playing career consisting of one at-bat—in which he struck out—had managed in the minors for several years but had never even served as a major league coach when he took over the Dodgers. His inexperience, injuries to key players, and an outstanding season by the Giants dropped the Dodgers to second place in 1954. For most teams, ninety-two wins and a second-place finish would have been very respectable; but the Dodgers had won pennants in 1949, 1952, 1953, lost a fourth flag in 1950 on the last day of the regular season, and, had it not been for "the home run," would have claimed yet another title in 1951.

Many saw Alston as a caretaker who would serve a couple of years until Pee Wee Reese, the Dodger shortstop and captain, was ready to retire and take over managing duties.[2] It had been assumed that Reese could have succeeded Dressen if he had wanted to be player-manager, but he preferred to concentrate on playing.

In 1955 Reese was thirty-seven years old and the only veteran of the 1941 pennant winners. Robinson was thirty-six, though he was entering only his ninth season, his career postponed by the color barrier he had eventually broken. Roy Campanella was thirty-four, ancient for a catcher. Carl Furillo in right field was thirty-three. Gil Hodges at thirty-one and Duke Snider at twenty-nine were comparative youngsters. With so old a team, the Dodgers' pennant chances were jeopardized, and they could expect strong challenges from the defending champion Giants and the Milwaukee Braves, who had added a promising rookie, Henry Aaron, to their young regulars. Nor could the Cincinnati Reds be ignored, with a lineup that included Ted Kluszewski, Frank Robinson, Gus Bell, and Wally Post.

After a one-day rainout, the Dodgers opened the 1955 season on April 13 against the Pittsburgh Pirates. John Cashmore threw out the first ball in place of Governor Averell Harriman, and a crowd of only 6,999 braved cold and damp conditions to see if Carl Erskine could extend the Dodgers' streak of three consecutive opening-day wins.[3] Erskine and Max Surkont of the Pirates pitched shutouts until the sixth inning, when the Dodgers managed four hits, including two doubles, but scored only once. The Pirates tied the game in the top of the seventh, but in the bottom of that inning the Dodgers secured the victory with a five-run rally that led to Erskine's first opening-day win in four attempts. Jim Gilliam, the second baseman, and Carl Furillo hit home runs, Duke Snider went three for three, and the infield turned three double plays.

The next day the Dodgers traveled across town to meet the Giants in their home opener at the Polo Grounds. Sal Maglie opposed Don Newcombe, who had finished 9–8 in 1954 after a twenty-win season in 1951, his last year before military service. The weather remained dismal, but ceremonies in honor of the Giants' world championship brought more than 29,000 to the park. The Giants broke on top with a run in the second, but the Dodgers chased Maglie with four in the fourth, including a three-run homer by Roy Campanella and solo shot by Newcombe. The Dodgers added a run in the fifth and five more in the seventh, including another homer by Newcombe and Carl Furillo's second round-tripper in as many games. The Giants rallied in the late innings, but Snider's brilliant catch of a Monte Irvin drive in the ninth preserved the win.

In spite of their two victories, the Dodgers remained half a game behind the Chicago Cubs, who were also undefeated but had played one more game. The Dodgers caught the Cubs with a 6–3 win over the Giants, sweeping the two-game interborough series. Furillo continued his slugging with two more home runs, and Billy Loes pitched a complete game. The Dodgers moved on to Pittsburgh, where Russ "Mad Monk" Meyer shut out the Pirates on two hits for a 6–0 win. The streak continued during a Sunday doubleheader. The Dodgers won the opener 10–3, as Johnny Podres went the distance, and Clem Labine won the nightcap 3–2. The Dodgers made it seven as Duke Snider hit a three-run homer behind Carl Erskine's pitching for a 5–2 victory over the Phillies in Philadelphia. In the next game the Dodgers' bats exploded for three homers, including Carl Furillo's fifth of the young season. The Phillies mounted a late rally, but fell short, losing to Newcome 7–6.

On April 21, the Dodgers tied a National League record by winning their ninth consecutive game since opening day. The victory came in Ebbets Field against the Phillies, the Dodgers rallying for three runs with two out in the seventh to overcome a two-run deficit. The next evening the Dodgers went for the record against the Phillies' ace, Robin Roberts, before only 3,874 fans. Four Dodger homers combined with fine pitching by Joe Black secured a 14–4 victory. The streak finally ended the next night when the Giants scored five times in the eighth to defeat Johnny Podres 5–4. In those ten games, the Dodgers opened a three-game lead over second-place Milwaukee and a six-and-a-half game spread over the Giants. More significantly, the fast start solidified Walter Alston's position as manager. Alston had provoked some controversy by batting Roy Campanella in the eighth position. Campanella, the league's Most Valuable Player in 1951 and 1953, had made his displeasure known to the press; if the Dodgers had stumbled out of the gate, serious problems might have developed in the clubhouse. As it turned out, Campanella became Alston's biggest fan, singing his praises after the record-setting win: "You've got to give top credit to the skipper. He's got real close to the players in the last few weeks and has done a great job."[4] Campy himself had also been impressive, leading the team with a .314 average. His 32 homers, 107 runs batted in, and .318 average would win him a third Most Valuable Player award.

The one flat note was the team's attendance. Only two of the games were played at Ebbets Field, and for both the weather was poor. Yet opening day and the record-setting tenth contest attracted fewer than 7,000 fans, combined. Only the interborough rivalry remained a potent draw, as more than 27,000 attended the eleventh game of the season, when the streak ended against the Giants.

Carl Erskine put the Dodgers back on the winning track with his third victory of the year, a 3–1 win over the Giants. The final game of the series was a customary Dodger-Giant contest. The Dodgers took a 1–0 lead in the bottom of the third inning on a home run by Pee Wee Reese. Willie Mays put the Giants ahead with a two-run shot in the fourth. The Dodgers replied with three runs in the bottom of the inning, aided by Carl Furillo's home run. Mays blasted another in the sixth to close the Dodger lead to 4–3. The Dodgers added a run in the seventh, but Mays kept the Giants in the contest by throwing Don Hoak out at the plate in the eighth inning. With one out in the top of the ninth, Dodger pitcher Billy Loes threw wide past first base, allowing Whitey Lockman to reach safely. Alvin Dark then tied the game with a home run. In the top of the tenth, the Giants seemed to put the game out of reach with six runs, aided by Furillo, who allowed a run to tag up and score when he caught what he mistakenly thought was the third out and neglected to use his strong arm to hold the runner at third. The Dodgers retaliated with five runs in the bottom of the inning but Furillo stranded the tying and winning runs.

Despite losing the series to the Giants, the Dodgers retained a two-and-a-half-game lead over Milwaukee. Brooklyn then embarked on another winning streak, which put the pennant out of reach. After a rainout, Billy Loes defeated Cincinnati 3–2, allowing only two hits. Forty-degree weather was partly responsible for the attendance plummeting to 6,032 after capacity crowds for the Giants series. Against the Cubs, Carl Furillo returned to form with his seventh home run, pacing a 4–2 win. By this time, Alston had moved Roy Campanella up to the clean-up spot, and Furillo now batted eighth, continuing to provide strong support from that normally weak position. Chicago fell again, 7–5, as homers by Gil Hodges, Jackie Robinson, and Duke Snider carried Russ Meyer to his seventeenth consecutive victory over the Cubs.

In their seventeenth game of the season, the Dodgers finally faced the second-place Braves, defeating them 5–4 on their first pinch hit of the year, a single by George Shuba. The loss dropped the Braves six and a half games off the pace, and they fell another half game back on May 3 when they lost to Pittsburgh while the Dodgers were idle. When Brooklyn beat the St. Louis Cardinals 4–3 for their seventh straight win, the Cubs moved into second place, seven and a half games back. In the midst of this prosperity, Don Newcombe provided some controversy by refusing to pitch batting practice as a protest against being used infrequently. He was suspended, but only for a day, as Alston inserted him in the eleventh inning of a game against the Phillies, which the Dodgers won 6–4. The front office modified Newcombe's penalty to the loss of a day's pay, which in those days before free agency, came to approximately $104.17.[5]

On May 7, Carl Erskine won his fifth game and the twentieth in twenty-two for the Dodgers with a 6–3 victory over the Phillies. The Giants and Braves were tied for second place, each playing .500 ball and nine games behind Brooklyn. The Dodgers completed a sweep of the Phillies with a 9–8 win on May 8. Hodges hit his fourth homer of the year, and Duke Snider hit the fourth grand slam of his career, helping Johnny Podres gain his third victory. The team then moved to Chicago, where Newcombe emerged from the doghouse to pitch the game of his career, a one-hit shutout in which he walked none and faced the minimum twenty-seven batters. An otherwise perfect game was marred by Gene Baker, the Cubs' second baseman, who lined a clean single to center with one out in the fourth. Baker was then thrown out trying to steal second. Snider's ninth homer of the year gave Newcombe the only run he needed in a 3–0 triumph.

The eleven-game winning streak seemed secure as the Dodgers sent Russ Meyer to the mound in pursuit of his eighteenth straight win over the Cubs, but Ernie Banks hit a grand slam in the first inning to lead the Cubs to a 10–8 victory. The loss was Brooklyn's third in their first twenty-five games, and left them leading the Giants by eight and a half, with Milwaukee in third place, nine back. Through May and June, the Dodgers increased their margin over their competitors as they overcame a number of injuries to open a thirteen-game lead over the Braves on July 1. The gap closed to eleven and a half on the Fourth of July, and remained there through the All-Star Game on July 12.

Despite dominating the National League, the Dodgers placed only three players on the All-Star team. Snider, who received a plurality of the fans' votes, started in center field, Gil Hodges was named as a reserve. Newcombe was bypassed as the starting pitcher by manager Leo Durocher despite his spectacular 14–1 record.

During the next month, the Dodgers put the National League pennant away, lengthening their lead over the Braves to fifteen and a half games on August 15. A season that had begun in doubt and had been plagued by injuries was on the verge of rewarding them with unprecedented success. Never before had the Dodgers so dominated the league, and at the peak of this dominance came O'Malley's announcement of the Jersey City schedule, the first clear sign of how fragile Brooklyn's hold on the team was. While machinations over the proposed new stadium developed, the ball club continued to hold a huge margin over the rest of the league, clinching their eleventh pennant since 1890 with a 10–2 defeat of the Braves in Milwaukee on September 8. This was the earliest a pennant had been decided in modern National League history, and credit for the win went to Karl Spooner, who, after his brilliant debut at the end of the 1954 season, had struggled with injuries in gaining an 8–5 record in 1955. For once the Dodgers were able to relax while the Yankees labored to wrest the pennant from the defending champion Indians.

In addition to Roy Campanella, other Dodgers turned in outstanding efforts. Carl Furillo finished at .314 with 26 home runs and Duke Snider hit .308 with 42 homers and a league-leading 136 runs batted in. Reese and Hodges hit .282 and .289, respectively, and Hodges added 27 homers and 102 runs batted in. Jackie Robinson, in his next-to-last season, fell off to .256 but continued to provide leadership.[6] Perhaps most remarkably, Don Newcombe, who began the season as a spot reliever, finished with 20 wins, second in that category only to Robin Roberts, who led the league with 23. Moreover, Newcombe actually had the highest batting average on the team, hitting .359 in 117 at bats. He also hit 7 home runs and on occasion was employed as a pinch hitter. The team retains a reputation as one of the all-time great clubs, in the same class as the 1927 Yankees. That season the Dodgers scored more runs than any other team, hit more doubles and home runs, had a higher batting and slugging average, and stole more bases. The pitchers were equally dominant, leading the league in fewest runs allowed, strikeouts, saves, and earned run average.

In 1952 and 1953, Brooklyn had entered the World Series against the Yankees after performances nearly comparable to that of 1955. In 1952 the Dodgers won the first game of the series, then held advantages of

two games to one and three to two before the Yankees edged them in
Ebbets Field 3–2 in game six and 4–2 in the decisive seventh game. In
1953 the team that had won 105 games during the season dropped the
first pair of Series games, tied the Yankees after taking the next pair,
and then lost games five and six and the Series. The 1952 and 1953
World Championships were the fourth and fifth consecutive titles for
the Yankees, a record that may never be equaled.

In 1955 Brooklyn would again face the same tormentors: Berra, Mantle, Martin, Rizzuto, and associates. The Series began on September 28
in Yankee Stadium, with Don Newcombe opposing Whitey Ford. The
Yankees prevailed, 6–5, despite home runs by Furillo and Snider and
a steal of home by Jackie Robinson. The second game was even more
disturbing as Yankee pitcher Tommy Byrne singled in two runs in the
fourth inning and became the first left-handed pitcher to hurl a complete game victory over the Dodgers in 1955. The defeat placed Brooklyn in a position from which no team had ever recovered to win the
Series.

The Dodgers' power returned in the third game in Ebbets Field as
they won 8–3 behind Johnny Podres, 9–10 on the year, who pitched a
complete game on his twenty-third birthday. In the fourth game, home
runs by Campanella (his second in two days), Snider (his second of the
Series), and Hodges paced an 8–5 victory, tying the Series at two games.
Snider hit two more home runs in game five as the Dodgers won 5–3
for Roger Craig, who had relief help from Clem Labine to sweep the
three-game set in Brooklyn. This remarkable revival merely restored
the Dodgers to the position they had been in three years before when
the Yankees had won games six and seven at Ebbets Field. This time
the Yankees had the advantage of returning to their home park in the
Bronx.

In the first inning of the sixth game, the Yankees scored five runs,
knocking Karl Spooner from the game after a third of an inning in what
would be his last appearance for the Dodgers. Whitey Ford stifled the
Dodgers on four hits and one run, becoming the second left-hander to
win a complete game against the Dodgers. Brooklyn's lineup, which
had devoured left-handed pitching all season, was stymied in Yankee
Stadium, where the power alley in left-center field was appropriately
named Death Valley. Before the renovations of the 1970s the dimensions of the Stadium were 301 feet down the left field line, 296 feet to
right, and 461 feet to the power alley in left-center. In those most un-
friendly confines, the Dodgers faced the task of beating Tommy Byrne
in the seventh and deciding game.

Walter Alston selected Podres to pitch, bypassing Newcombe and

Erskine who both were ready and who had had better records than Podres during the season. Neither Newcombe nor Erskine had pitched effectively in the Series, however, and both were right-handers in a Series dominated by southpaws. The final game was indeed a pitchers' duel. The Dodgers scored two runs, both driven in by Hodges. The Yankees threatened in the sixth when Berra sliced a drive down the left-field line with two on and one out. Sandy Amoros's catch is perhaps the fielding equivalent of Thomson's home run; every fan sees the picture of the left-fielder straining with his gloved right hand to snare Berra's drive. The knowledgeable fan will also mention that Alston sent Amoros in that inning to replace Jim Gilliam, who as a right-hander would have had to backhand the ball, a very difficult chance. Peter Golenbock recounts that to the players the notable feature of the play was not Amoros's catch but the throw that followed, a strike to the ever-alert Pee Wee Reese was relayed to Gil Hodges, doubling up Gil MacDougal to end the inning and stifle the Yankee threat.[7] Again in the eighth, the Yankees put the tying runs on with one out, but Podres got Berra and Bauer, ending the inning and the final Yankee rally. In the ninth, Podres retired Bill Skowron on a bouncer back to the mound for the first out. Bob Cerv then flied to left field. Elston Howard grounded a 2–2 pitch to Pee Wee Reese at shortstop, and Reese threw low to Hodges, who handled the throw for the Dodgers' first world championship in eight tries.

The scene in Brooklyn was comparable to V-E Day. The Dodgers had not only won a championship; they had beaten the tag that they folded under pressure. They also restored some pride to a borough that had been losing status through most of the century. Unfortunately, even in this triumphant moment the signs of Brooklyn's doom were apparent. After the Dodgers' previous World Series losses, the *Brooklyn Eagle* had run as a headline the famous cry, "Wait 'till Next Year!" On this grand occasion, "This Is Next Year!" graced the front page of the *Daily News;* the *Eagle* had folded ten months before.

The victory naturally obscured critical problems facing the team in Brooklyn. Who could seriously consider that the Dodgers would leave Brooklyn *now,* after an achievement that had eluded them for more than half a century? People may have overlooked that just prior to the Series that Dodgers announced they had retained Buckminster Fuller to provide studies for a new domed stadium in Brooklyn.[8]

The months after the World Series brought additional joy to the Dodger offices on Montague Street, in Brooklyn Heights. In the not inconsequential matter of net profit after taxes, the Dodgers had been the most successful team in baseball. Their net income for 1955 was

$427,195, ahead of the Braves, who took in $409,023.[9] These figures are especially important because it was the Braves' extraordinary burst in revenues when they moved to Milwaukee that stimulated O'Malley into forcing action on a replacement for Ebbets Field. In their last year in Boston, the Braves lost $459,099; their first year in Milwaukee they turned a profit of $637,798. That sum dwarfed the Dodgers' net income of $290,006 for 1953, even though the Dodgers won the pennant that year and enjoyed revenue from a six-game World Series.

In 1954 the Dodgers' profit slid, but many teams would have been delighted at finishing over $200,000 in the black, a return that alarmed O'Malley because the Braves posted a profit of $457,110. The continued push for a new stadium in light of the Dodgers' financial success is perhaps the most galling point to those who blame O'Malley's greed for the Dodgers' eventual move. The other teams that changed cities in the 1950s were destitute franchises. In addition to the anemic Braves, the Browns lost over $1 million in their last two years in St. Louis, and the Athletics lost over $350,000 in their last three seasons in Philadelphia.[10] For those teams moving may have been their only means of survival.

The Dodgers, on the other hand, were one of the most profitable teams in the major leagues and the only National League team to make money each year from 1952 to 1956. Their total profit for those five years exceeded $1.7 million, surpassing the net income of even the Yankees.

The Dodger owner may have been easy for the public to hate, but the franchise's financial picture in the mid-1950s is misleading. The team had dominated the league in that decade as few clubs ever had, and the financial returns were to a great degree a result of that success; but O'Malley was aware that the Boys of Summer were aging, and that some lean years were probably inevitable as the club rebuilt. He had to anticipate what revenues would be once the team became less of an attraction. In that regard, a closer examination of Brooklyn attendance is useful. For all the passionate prose written about the devotion of the Flatbush Faithful, the attendance figures are rather underwhelming. In their world championship season, the Dodgers attracted 1,033,589 fans, an average of 14,355 per game, about half the capacity of Ebbets Field.[11] The Dodgers' park may have been lovably old, dirty, and decrepit, but it was not jammed to the rafters with cheering partisans.

Some mitigating factors should be noted. First, although the season did not begin as early in April as is currently the case, the first games were often played during chilly evenings. Second, the structure of Ebbets Field included a number of vantages that obstructed the fans' views

with pillars and girders. Finally, televised games cut into daily attendance. Still, one can't help concluding that the rosy picture of Brooklyn
as a persistently supportive community is somewhat deceptive. Given
the demographic shifts of the borough during the 1950s, one could
hardly have expected a business so dependent on the support of the
general public to be complacent about what could well prove to be
fleeting success. Annual shifts in Brooklyn's attendance and revenues
were sizable during those years:

	Attendance[12]	Net Income[13]
1952	1,088,704	$446,102*
1953	1,163,419	$290,006**
1954	1,020,531	$209,979
1955	1,033,589	$427,195*
1956	1,213,562	$487,462*

*includes revenues from seven World Series games.
**includes revenues from six World Series games.

For teams less successful than the Dodgers, net income could fluc-
tuate even more wildly. In 1952 the Chicago Cubs profited by $150,000,
and then lost more than $400,000 the following year.[14] The St. Louis
Cardinals lost over $40,000 in 1955 and then cleared over $300,000 in
1956. Baseball accounting is a notoriously uncertain pursuit, since many
teams are owned by individuals whose major financial interests lie in
other businesses. Accordingly, a franchise may not be in quite the condition it appears to be on paper. For example, from 1952 to 1956, Branch
Rickey served as general manager of the Pittsburgh Pirates, lost almost
$2 million during that stretch.[15] These were the most serious losses
sustained by any major league franchise at that time, but they reflected
in part Rickey's investment in a farm system that led to the Pirates'
world championship in 1960 and their consistent role as a National
League power for many years after.

On the face of it the Dodgers recorded net revenues in the mid-1950s
beyond those of any other major club league, but those years represented a high-water mark for the Dodgers in the Ebbets Field era, and
the impact of demographic and cultural change that had already driven
some teams to new cities would inevitably force the Dodgers to make
adjustments as well. To Walter O'Malley the most crucial adjustment
was a new playing facility that could extend modern conveniences appropriate to the postwar era.

The constraints of Ebbets Field had prompted speculation earlier in
the 1955 season that if the Dodgers and Yankees met again in the World

Series all the games might be played in Yankee Stadium.[16] The six-game 1953 series included three games in each of the two home parks. The games in Yankee Stadium attracted 198,520 fans and yielded receipts of $1,141,098.10.[17] The games in Ebbets Field drew only 108,820, with a take of $638,171.34. O'Malley emphasized that borough attachments required that the Dodgers' home games in the 1955 World Series be held in their own park, but civic pride has its limits, especially when nearly $200,000 per game is being sacrificed.

Another concern for O'Malley was the club's cost of operations. The Dodgers' success in the 1950s was a direct result of investments made by MacPhail, Rickey, and O'Malley. The farm system and the spring training facility at Vero Beach, Florida, were among the most elaborate in baseball. One estimate placed the cost of spring training at $200,000.[18] Such expenses could keep a club in the red unless a firm revenue base was also maintained.

By December 1955 some of the broader implications of the Dodgers' scheduling games in Jersey City were becoming clearer. Writing in the *Sporting News*, columnist Dan Daniels predicted the Dodgers would probably sell Ebbets Field in the near future and would need a new playing facility by 1958. He anticipated that if a new stadium was not ready in Brooklyn by that time the Dodgers might move not just to New Jersey but break all the way to Los Angeles. Chiding Brooklyn for the lack of support for the world champions, Daniels wrote, "Brooklyn fans have developed a strange mental attitude toward their ball club. The customers have an idea that the club is a public enterprise and that its owners are forced to be public benefactors."[19] Another New York writer, Roscoe McGowen, wrote that the major effect of the Jersey City move was to wake people up to the fact that baseball in Ebbets Field was in its final seasons: "Unless O'Malley succeeds in getting the site he wants—at Flatbush and Atlantic Avenues—for his new stadium, it is as definite right now as anything can be that there will be no stadium and no Dodgers in Brooklyn."[20]

At the same time, O'Malley issued a press release denying that the Dodgers had any intention of moving permanently to Jersey City. He described the scheduling of some games in Roosevelt Stadium as being "in furtherance of the fine co-operation we have had from the Brooklyn borough president and the Board of Estimate and is a long range safeguard to protect the franchise for Brooklyn—and nothing more."[21] O'Malley announced also that "we fully appreciate that it is unlikely that progress can be sufficiently rapid to have our new Brooklyn stadium available for the 1958 season, in which event our present arrangement with Jersey City can be viewed as guaranteeing the continuance of the franchise at the nearest possible point to Brooklyn during the period when the new stadium might be in the course of construction."[22]

In that curious time of basking in their first world championship and counting unprecedented revenues, the Dodgers were establishing limits that would within three years force either a new stadium in Brooklyn or else their move from the borough itself. Even those who could see the paradox of the winter of 1955 could not always see its full implications. For instance McGowen wrote, "O'Malley's statement means that with a new Brooklyn stadium not 'available' when Ebbets Field is abandoned at the end of 1957, the entire season's home schedule of 1958 will be played in Jersey City. Maybe it also means 1959, and 1960, 1961 and. . . ?"[23] In McGowen's worst case, the Dodgers would move from an inadequate stadium in Brooklyn to an inadequate stadium in New Jersey, and they would stay in the area of greater New York until the stadium of their dreams appeared.

Specific steps to develop a new stadium were finally undertaken in February 1956 when Mayor Robert Wagner sent a bill to the state legislature for the purpose of establishing a Brooklyn Sports Center Authority. This agency would be empowered to transfer land in the manner of the Housing Act's Title I powers, and it also could sell bonds to construct a publicly owned sports stadium.

Although Wagner ostensibly sent the bill to the state capital, he stopped short of endorsing the measure. The bill's passage was in serious doubt, for several reasons. First, public authorities were being challenged for what many considered their excessive autonomy, and proposals were being considered to reduce the number of authorities already in existence. Second, the ability of the authority to sell its $30 million in bonds was generally questioned. The Dodgers had announced that they would purchase $4 million themselves, but the balance of the market was far less certain. Third, by redeveloping 110 city blocks for a public purpose with public funds, the state was removing a substantial source of tax revenues from the books. Inevitably those revenues would have to be made up elsewhere, and potential targets in the other boroughs were quick to see the threat. Finally, there was the nemesis of constitutionality. The transfer of land from one private party to another remained a controversial proposition which a number of groups threatened to test in the courts should the Sports Center Authority being considered by the state legislature in 1956 be established.

A procedural impediment also imperiled the bill. A state law being proposed to aid a single city meant that special protections for the rest of the state's citizens must be observed. The bill had to be specifically requested by the mayor of New York with a majority vote of the city council or, in the absence of the mayor's support, the request must be made by a two-thirds majority of the city council. Then the bill must pass each house of the state legislature with a two-thirds majority. Because Mayor Wagner had not lent his backing to the proposal, a two-thirds vote was needed from the twenty-five member New York City Council, comprised of twenty-three Democrats and two Republicans, each borough represented as follows: Brooklyn—9; Bronx—5; Manhattan—6; Queens—4; Staten Island—1.[24]

Although these formidable political and governmental barriers threatened the new stadium, another obstacle had been removed. Walter O'Malley announced that while his preference was to construct his own stadium with private funding, he would be willing to have the Dodgers play in a state-owned facility as tenants of the Sports Center Authority. This accommodation suggests that the Dodger president was receptive to a number of proposals for a new stadium as long as the site remained in Brooklyn. Hopes for the legislation were buoyed further by the announcement that its passage was endorsed by Robert Moses, who apparently was more sympathetic to this method because the state would retain ownership of the facility, an outcome not possible if the land had been transferred to the Dodgers under federal Title I powers.[25]

Other conflicting voices ensured that the bill faced an uncertain future. The Brooklyn Chamber of Commerce endorsed the measure; other groups threatened law suits if it passed. State legislators expressed concern that a precedent might be established for other sports franchises; The New York Times lent qualified editorial support, so long as the ballpark was "only incidental, fractional to a larger purpose."[26] Arthur Daley in his Times column stressed the importance of the legislation by noting that "the last thing that O'Malley wants to do is have the Dodgers leave Brooklyn. It would be a move born of desperation. If the Sports Center Authority is not the answer, another answer must be found."[27]

The Authority cleared its first hurdle on the last day of February when the city council, after several failed attempts, voted 17–5 with 3 absent to pass a Home Rule bill that enabled the legislation to be considered in Albany. The mayor's coolness toward the measure was underscored several days later when he reported that he was "not married" to the bill.[28] Since the mayor needed to sign the bill even if it passed and was signed by the governor, Wagner's aloofness signaled potential problems for the entire legislative effort. Republicans in the legislature were equally unsympathetic to the bill, but were reluctant to appear to be the cause of the Dodgers' move from Brooklyn. Accordingly, they offered little opposition, hoping to put Governor Harriman and Mayor Wagner on the spot. As these machinations progressed, Moses cooled his earlier endorsement by cautioning people not to "take this business of 500 acres too seriously."[29] He indicated that while a park for the Dodgers might prove feasible, the redevelopment of the entire 110-block site proposed by the statute might be too ambitious. Moses reminded reporters that the bill was only an enabling act, and that the task of selling the bonds for the project might prove impossible.

The curious nature of the bill was demonstrated again on March 19 when the state assembly voted passage 114–32, well over the two-thirds majority required. In its coverage of the vote, the *Times* suggested that the margin of victory might have been secured by the rather anonymous device of calling for a show of hands rather than the alternative of roll call.[30] The *Times* noted that observers had speculated that the necessary votes might not have been forthcoming if the roll call had been invoked. Two days later the senate passed the measure, 40–17. Another regional alternative to Ebbets Field was offered by Senator Edward Curry, a Democrat from Staten Island who offered that borough as a new home for the Dodgers.

In its report on the new bill, the Citizens Budget Commission expressed approval of the bill's protections for the city. Two features were specifically lauded. First, expenditures could come only from the Authority's revenues. If selling the bonds proved unfeasible, then the Authority couldn't borrow additional revenues to finance the construction of the stadium. Second, any plans for redeveloping the area were subject to prior approval by both the City Planning Commission and the Board of Estimate.[31] In many ways, these checks addressed the concerns of those who feared the Authority's autonomy but they also compromised its effectiveness.

In the days before the 1956 season opened, Robert Moses reiterated his warning that the redevelopment project was unrealistic. He advocated a more modest plan in which the new stadium would be only a single component.[32]

As one reviews the objections posed by various critics the outlines of a dilemma emerge. Moses claimed to support the bill, but routinely cautioned against excessively elaborate plans. His warning specifically was that the more complex the redevelopment plans, the more difficult they would be to effect, especially in terms of selling the bonds. On the other hand, a criticism clearly expressed by many, and implied by the *Times*'s editorials, was that the entire Sports Center Authority needed to be more than an elaborate ruse to construct a new stadium for the Dodgers. As the call for a larger public purpose grew louder it became increasingly difficult to meet Moses's objections. Thus, the new stadium had to be a minor element in a design whose grandiosity ignored financial realities.

On April 19, 1956, the Dodgers brought major league baseball to Jersey City, defeating the Phillies 5–4 before more than 12,000 fans who braved icy temperatures. Three days later, Governor Harriman went to Brooklyn to sign the Brooklyn Sports Center Authority Act, leaving to Mayor Wagner the final responsibility of selecting the members for the non-salaried positions. One month later, the *Times* carried a story indicating that Wagner intended to make the appointments within the week.[33] May was propitious for the Dodgers, as Carl Erskine pitched a no-hitter against the Giants, becoming the first Dodger hurler to accomplish the feat twice. In a dramatic shakeup of the pitching staff, the front office sold Billy Loes to Baltimore and replaced him with Sal Maglie, their former nemesis with the Giants.

These encouraging on-field developments were not matched in the political arena. May passed without the heralded appointments to the Sports Center Authority. It was now taking as long to name the members of the Authority as it had taken to establish the agency through the legislative process. In the last week of June, the engineering firm of Clarke and Rapuano filed an interim report with the Board of Estimate, indicating that a new ballpark could indeed be constructed on the Atlantic-Flatbush site.[34] This report, funded by the Board in 1955 shortly after Walter O'Malley announced the scheduling of games in Jersey City, was a preliminary evaluation of the redevelopment proposal, and it indicated the need for further consideration of more specific plans.

The halfway point of the season passed with the Dodgers locked in a fierce battle with Milwaukee and Cincinnati, and with the Sports Center Authority still existing only on paper. On July 23, the *Times* published a story with ominous findings: the redevelopment of the Atlantic-Flatbush site would remove property from the city's tax rolls that had been assessed at nearly $110 million.[35] The expense in lost revenues was estimated at $5 million. An autonomous city could have decided that the investment was sound, not only because the Dodgers would have had a new stadium, but because the redevelopment of a deteriorating neighborhood might have stimulated financial growth. But Brooklyn had to secure the cooperation of the other boroughs, whose representatives on the Board of Estimate would no doubt wonder how the $5 million was to be made up. And the mayor still had not appointed the Authority members.

On July 25 the *Times* reported that the overall cost of redeveloping the site would be $250 million.[36] The location for the new stadium was to be bounded by Warren Street, Fourth Avenue, Flatbush Avenue, and Fifth Avenue. Further progress was suggested by the mayor's announcement that Charles Mylod, Robert Blum, and Chester Allen would serve as the Board of the Sports Center Authority. Even this announcement was greeted with formidable opposition. O'Malley, Moses, and officials of the Long Island Rail Road expressed their disappointment over the site, indicating that they preferred Hanson Place, Fort Greene Place, Atlantic, and Flatbush avenues.[37] As Moses pointed out, the latter site was in the truly depressed area, while the space recommended to the Board of Estimate was far more valuable land.

The season entered its final phase with no further progress on a replacement for Ebbets Field and with the Dodgers tied with the Braves. Brooklyn lost three of four games to Pittsburgh and dropped one behind Milwaukee with just five to play. The Dodgers received a big boost from Sal Maglie, who no-hit the Phillies on September 25, but Milwaukee kept pace with Warren Spahn's twentieth win and preserved their half-game lead. Robin Roberts defeated Don Newcombe in the following game to put the idle Braves a full game ahead of the Dodgers heading into the final series of the season.

While the Dodgers entertained Pittsburgh at home, the Braves were on the road against the Cardinals. The Dodgers gained half a game on Friday night—they were rained out—while the Braves lost to St. Louis. Saturday was pivotal, as the Braves suffered a bitter loss, 2–1, in twelve innings and the Dodgers swept a double-header from the Pirates, 6–2 and 3–1. The victories vaulted the Dodgers into a one-game lead over the Braves, who could now hope only for a playoff if fortune prevailed on the final day of the season. The Braves finally did their part, beating the Cardinals 4–2, but the Dodgers secured their final pennant in Brooklyn with an exciting 8–6 win. Experience was perhaps the decisive factor; the Dodgers won when they had to, while the young Milwaukee team stumbled against the Cardinals, who finished below .500.

In the postgame joy, Walter O'Malley announced that "real progress" was being made toward a new stadium, although a closer look at those final games would have provided additional cause for concern.[38] The Dodgers played their last five contests in Ebbets Field, beginning that sequence in second place by half a game. Maglie led off with his no-hitter, hurled before 15,204 fans. On Thursday night, when one would have expected a large crowd in the wake of the no-hitter and with the pennant on the line, the Dodgers drew only 7,847. "Is that civic pride?" Arthur Daley asked the next day in his column. "Is Brooklyn really the wackiest and most fanatical baseball town in the country?"[39] Daley concluded his piece on "The Deserted Village" by noting that the new domed stadium would merely provide more seats to stay away from. A more appropriate crowd of 34,022 cheered the Dodgers during their Saturday sweep of the doubleheader, and another 31,983 were in attendance as the Dodgers clinched the flag. Ebbets Field was hardly a ghost town; but when weeknight crowds during the climax of a pennant race filled less than half the park, the Dodgers could hardly rely on the unstinting support of the borough in the future. Certainly, television was partly responsible for the decisive game of the season being played before some empty seats. One can also assume that Maglie's no-hitter was witnessed by thousands more than the modest numbers who viewed the achievement in person. This explanation, however, does not solve the problem of how the Dodgers could continue to attract numbers in Brooklyn sufficient to guarantee the revenues needed for the team to remain competitive.

The World Series, which began on October 3, provided the first clear sign that the era of the Boys of Summer was passing. The seventh meeting between the Dodgers and the Yankees opened in impressive fashion for the Dodgers. They won the opener 6–3 in Ebbets Field behind Sal Maglie's complete-game victory over Whitey Ford. Gil Hodges and Jackie Robinson homered, overcoming a two-run blast by Mickey Mantle in the first inning. The second game was a wild affair. The Yankees jumped out to a 6–0 lead, paced by Yogi Berra's grand slam home run in the second inning. The Dodgers rallied to tie in the bottom of the inning, powered by a three-run homer by Duke Snider. Both starting pitchers, Don Newcombe and Don Larsen, were chased in the second inning, and Don Bessent came on for the Dodgers to pick up a 13–8 win in the longest nine-inning game in Series history. The victory put the Dodgers in the same commanding position the Yankees had been in the year before but, rather than the prelude to another world championship, victory proved to be the last hurrah of the Dodger sluggers.

Back in Yankee Stadium, Whitey Ford regained his form, beating Brooklyn 5–3. The Yankees then tied the Series at two games apiece with a 6–2 triumph in game four. The pivotal fifth game brought Sal Maglie back to the mound fresh from a no-hitter and a win in the opener. But this game belonged to Don Larsen, who hurled a perfect game, the only no-hitter in Series history. The game in fact was reasonably close, since Maglie didn't allow a hit until Mantle homered in the fifth inning. Larsen was aided by great fielding, and he received an insurance run in the seventh inning. Yankee pitchers continued their mastery in the sixth game as Bob Turley permitted the Dodgers only four hits. One of the blows, however, was a single by Jackie Robinson in the tenth inning, and that proved to be the winning run, as Clem Labine shut out the Yanks in Ebbets Field to send the Series to a seventh game.

The Dodgers went into the finale in Ebbets Field with an offense on the point of futility; with twenty-seven-game-winner Don Newcombe on the mound, however, they stood a chance of repeating as world champions. But the Yankees drove Newcombe from the game in the fourth inning, and two home runs by Berra and a grand slam by Skowron led the Yankees to a 9–0 rout. In the final three games, the Dodgers had managed a total of one run on seven hits, a dismal performance and also the last appearance in the Series by a Brooklyn franchise.

On October 31, 1953, just three weeks after the embarrassing seventh-game defeat, and while the Brooklyn Sports Center Authority languished in administrative paralysis, Walter O'Malley sold Ebbets Field to developer Marvin Kratter.[40] The agreement called for the Dodgers to lease the stadium back for five years, though Kratter indicated they could use it for a longer period while the new stadium was being constructed. The new owner of the historic old park declared he "would never do anything to keep the Dodgers from playing in Brooklyn." O'Malley, however, had different ideas. When asked during his 1957 testimony before the House Antitrust Subcommittee if he could obtain an extension on his lease with Kratter, O'Malley replied, "When a businessman commits himself to pay $3 million for a piece of property, I don't think it is fair to try to renegotiate that deal after it has been made."[41] The Dodgers' president appears to have intended to bring the issue of a new stadium to a head and to force New York officials to decide whether they would cooperate in constructing a new stadium. By extending the time the Dodgers played in Ebbets Field, O'Malley would probably have merely invited Parkinson's Law, drawing out the decision-making process without a final determination becoming any more likely.

The immediate financial outlook for the Dodgers was even more attractive for the team in 1956 than it had been the preceding year. Despite the poor crowds for some crucial games down the stretch, the tight pennant race, in contrast to the 1955 romp, substantially increased attendance. The turnstiles at Ebbets Field admitted 1,213,561 fans in 1956, an increase of 179,973 from 1955.[42] Revenues from the

increased attendance jumped almost $250,000 to $1,790,275, and revenues from broadcasting grew by another $100,000 to $888,270. The
broadcast revenues were again critical because Milwaukee garnered only
$175,000 from that source in 1956, and the Braves' net income totaled
$362,268 compared to the Dodgers' $487,462.[43] These figures include
the revenue the Dodgers received from the seven-game World Series,
so the financial advantage over the Braves must be recognized as somewhat tenuous. Just as after the 1955 season, the Dodgers were in an
enviable position for the short run, but their time in Ebbets Field was
now officially limited and progress on an alternative site was very slow.

On November 20, 1956, the engineering firm of Clarke and Rapuano
issued a planning study for the area of Brooklyn that included the proposed new stadium.[44] The study included reports from a real estate
firm and from attorney John McGrath, who studied the legal basis for
the development. The study recommended that a fifty-thousand-seat
stadium be constructed "in the area bounded by Flatbush Avenue,
Fourth Avenue and Prospect Place." This site was across Atlantic Avenue from the Long Island Rail Road station, the same site O'Malley
and Moses had criticized several months earlier. Clarke and Rapuano
further proposed the construction of two parking garages to accommodate 2,500 cars, and they also wished to build a concourse to handle
the pedestrian traffic from the stadium to the railway terminal and the
subway lines. These recommendations were the most tangible indica-
tions so far that a stadium might yet be built to keep the Dodgers in
Brooklyn. The next step was for the Sports Center Authority to secure
sufficient funding from the Board of Estimate to conduct the necessary
follow-up studies. The Authority requested $278,000 for these studies,
and the Board considered the request in its December meetings.

On December 7, 1956, Robert Moses sent a memo to Mayor Wagner
apprising him of recent developments. In Moses's view, the Board was
reluctant to grant such a sizable sum to the Sports Center Authority.
Moses made two proposals to "overcome this stalemate. . . ."[45] The
first was to set up a Brooklyn Terminal Public Improvements Committee for the purpose of reviewing several matters pertaining to the Atlantic-Flatbush site, including resituating the Long Island Rail Road station in a location south of Atlantic Avenue, eliminating traffic congestion
in the area, and removing the meat market. The second proposal was
to provide $40,000 to the Authority to conduct a study on the economic
feasibility of the proposed stadium. This sum fell far short of what the
Authority had requested; Moses also advised that if bonds were eventually sold the Authority should repay the entire amount. The latter
point seems trivial on the face of it, but Moses noted in his memo that

Charles Mylod, the Authority's chairman, objected to any repayment plan, perhaps anticipating that a much larger sum would be forthcoming. Moses rejected Mylods's concerns as "sheer nonsense."[46]

Moses concluded by noting that the Authority did not need a large staff until it was demonstrated that the bonds could be sold. He advised the Sports Authority to meet with the Terminal Committee and that a report be issued by April 1, 1957. Moses made a final suggestion: "If the economic study by the Authority shows that the Sports Center is not practical, nothing will have been lost because it will then be possible for us to pick up the problem where the Sports Center left off and use the vacated property for the Title I project we recommended a year ago."[47]

The memo suggests that Moses's opposition to a new stadium remained implacable. His conclusion, first expressed in August 1955, that he had no confidence in such a scheme continued to guide the advice he offered Mayor Wagner. Moses's specific recommendations in December 1956 were to ask for one more study of the economic feasibility of the stadium, to establish yet another group to consider the redevelopment issues, and to provide the Sports Authority with a fraction of the money it had sought. If the Authority could not overcome those impediments, Moses was prepared to take over the project to construct the public housing project that he had been favoring for more than a year.

Whether Moses's recommendations decisively affected the events that followed is uncertain. He announced publicly on December 24 that he would construct Title I Housing if the Sports Authority could not sell its bonds, and four days later the Board of Estimate acted.[48] The stadium issue was cut off from the rest of the redevelopment project, which made the connection between the stadium and a larger public purpose even more tenuous. The Board also allocated $25,000 to the Sports Authority to consider the issue of economic feasiblity. This amount was less than one tenth what the Authority had sought, and was well below what even Moses had advised.

The Clarke-Rapuano report of November had indicated that progress toward keeping the Dodgers in Brooklyn might at last be possible and showed the way to a plausible alternative to Ebbets Field. The authors of the report anticipated that their "recommendations will meet with differences of opinion."[49] The action of the Board, however, simply derailed serious consideration of the site. The differences of opinion expected by the engineering report were not ironed out; they were evaded.

The Board's decision to award a negligible amount for additional study characterized New York's response to O'Malley's challenge. Those who wished to keep the Dodgers in Brooklyn faced the persistent opposition of Robert Moses, the hostility of officials from other boroughs, and the apparent indifference of the mayor. The efforts to overcome those forces were becoming increasingly futile.

5

Los Angeles: An Overture from the Coast

The political intransigence that greeted Walter O'Malley's efforts to construct a new stadium in Brooklyn afforded an opportunity to public officials in another city that sought to secure a major league franchise. In contrast to the unbudging opposition of Robert Moses and the relative indifference of Mayor Wagner, public officials in Los Angeles exhibited great determination and creativity as they guided the stadium issue past a set of obstacles even more complex than those New York had presented.

Los Angeles is commonly thought of as a postwar creation of cars, suburbs, and television. In fact, Los Angeles antedates the United States Constitution, having been chartered under the authority of Spain on September 4, 1781. Earlier yet, in the sixteenth century, Southern California was explored by the forces of Cortez and Coronado. Spanish sovereignty prevailed until 1821, when Mexico won its independence from Spain and established control over the province of California.[1] Before long, the commercial opportunities of California attracted Yankee traders, and led the United States to covet the land. The annexation of Texas in 1845 had demonstrated what American military pressure could wring from Mexico, and a similar method was employed to claim California for the United States in 1846. In his history of Los Angeles, John Weaver notes, "The Mexican provincial capital had become so Americanized that it outlawed bullfighting in 1860 and formed a club 'for the promotion of the manly art of base ball playing.' "[2]

Los Angeles developed as a significant urban center in the late nineteenth century, when sufficient water resources were harnessed to permit growth in a desert climate and effective transportation—namely, the Southern Pacific railroad—linked the city with other commercial markets. Water and the railroad proved mixed blessings, however, for the obvious benefits of these enterprises were tempered by the political corruption that infected them.[3] Throughout California, government served the interests of the Southern Pacific. The railroad dominated the political machines that controlled the public purse in Sacramento, San Francisco, and Los Angeles, just as machines raided the treasury in New York, where citizens suffered at the hands of their own railroads and of the Tweed Ring during these same years.

In New York the Progressive movement led to the expansion of the city through the merging of Brooklyn and Manhattan, which promoted economic efficiency and commercial growth and also created a federal system that vested primary municipal power in the Board of Estimate. Nonetheless, the more entrenched interests in Eastern politics were able to resist these democratic forces, and resume, albeit in a more chastened manner, the traditions of elite rule.

In California reforms were more sweeping. Suspicion of those in authority was so strong that organizational changes were given less emphasis than reforms facilitating the direct exercise of governing power by the electorate itself. In 1903, Los Angeles adopted three devices that provided citizens with the tools to formulate public policy, to change decisions already made by their representatives, and to remove office holders who had egregiously offended their constituents.[4] These reforms—called initiative, referendum, and recall—were mainstays of the Progressive movement and found their way into the government operations of many Western and Midwestern states.

By the early twentieth century, Los Angeles had become a complex community in which powerful monied interests nervously coexisted with Progressive Republicans in the Teddy Roosevelt tradition and also with Socialists who nearly got their candidate elected mayor in 1910. The Progressive forces (personified by Hiram Johnson, who was elected governor and later U.S. senator) emerged as a powerful political influence that guided the city between the reactionary politics of Harrison Gray Otis, publisher of the *Los Angeles Times*, and the Socialists, whose violent tendencies threatened to continue the frontier atmosphere of the nineteenth century.

Another contribution of the Progressives was to depoliticize the exercise of government. They rejected competition by interest groups for the largess of the state in favor of a dispassionate administration of government. As Robert Fogelson has described the Progressives in Los Angeles, "They also postulated, without defining, so fundamental an agreement about the goals of the metropolis that the functions of city government became administrative instead of legislative, a matter of executing rather than of determining policy."[5] This view of government, though perhaps naive, has persisted throughout the twentieth century. The Los Angeles mayor's office is a nonpartisan post customarily sought by several candidates, none of whom makes an issue of party affiliation. This approach contrasts with New York, where mayoral candidates are nominated by their parties and a candidate without partisan backing faces the difficult task of forging a fusion ticket.[6]

The political environment which Walter O'Malley faced in the 1950s was primarily characterized by a "strong council, weak mayor" form of municipal government. In New York legislative power was exercised, as we have seen, by the Board of Estimate. No such federated structure was established in Los Angeles, even though the city limits extended over four hundred and fifty square miles and encompassed many diverse local communities. The fifteen-member city council is elected from geographic regions, but none compares in governmental significance with the boroughs of New York.

The impact of the reform spirit in Los Angeles has been described by John Ries and John Kirlin:

> Originally adopted during January 1925—and true to the spirit and letter of the Progressive movement, whose reform agenda would exorcise not only the evils of the Southern Pacific Railroad (the "Octopus," as Frank Norris called it in his famous novel), but, also, corruption and spoils in government by banishing partisanship—the Los Angeles City Charter parcels out legal authority to more than a dozen boards or commissions. The Office of Mayor is almost treated as an afterthought.[7]

Another administrative characteristic of the Los Angeles government that would prove significant in the Dodgers' move is the role of the county government. Los Angeles County includes more than four thousand square miles, seventy-seven incorporated cities, and more residents than all but a few states in the union.[8] The county is run by a seven-member Board of Supervisors which oversees the services provided to the cities within its jurisdiction, as well as the vast unincorporated areas that have attracted sizable populations. The plan that Los Angeles ultimately offered to the Dodgers required the cooperation of city and county officials. Relations between these governments had not always been amicable, and there was a possibility that Los Angeles's bid could flounder under the same kinds of political forces plaguing New York.

The mayor who orchestrated Los Angeles's efforts to lure the Dodgers was Norris Poulson, elected in 1953. In his campaign he attacked the city's holding of 183 acres of land at Chavez Ravine for the purpose of constructing public housing.[9] Poulson criticized the expense of the endeavor, and benefited from popular sentiments associating public housing with socialism. The land had been acquired under the Federal Housing Act of 1949, the same act that Robert Moses had declined to invoke in Brooklyn on Walter O'Malley's behalf.

Chavez Ravine was named for one of the first councilmen of Los Angeles, Julian Chavez, who had moved to the pueblo from New Mexico in the 1840s. His land, just a few miles from the center of Los Angeles, was appraised in the 1850 census as worth $800.[10] A century later the land was still relatively undeveloped, but the passage of the Housing Act would change that. In October 1950, Los Angeles's mayor, Fletcher Bowron, unanimously supported by the city council, signed a contract to commit to the construction of ten thousand dwelling units at a cost of $110 million.[11] Almost immediately the agreement triggered bitter reaction from those who associated such public enterprise with socialism and communism.

At the same time Robert Moses was transforming New York by masterfully employing an array of governmental powers, Los Angeles was caught up with much of the rest of the country in a crusade against the threat of communism. A congressman from Southern California, Richard Nixon, rose to national prominence on the strength of his anti-Communist activities. National attention focused on the successful Communist revolution in China (which gave rise to the question, "Who lost China?"); on the Korean War; and on Wisconsin senator Joseph McCarthy's frantic pursuit of Communists in government. In Los Angeles the motion picture industry came under scrutiny, as some of its members were called before the United States House Committee on Un-American Activities to answer the question: "Are you now or have you ever been a member of the Communist party?"

In this environment, the housing projects at Chavez Ravine became a natural target. In his study of the controversy, Thomas Hines writes, "A cluster of groups, institutions, and individuals including the powerful California real estate lobby, the Home Builders Association, the Chamber of Commerce, and most critically, *The Los Angeles Times*, began to wage with increasing intensity an attack on the housing program as 'creeping socialism'—if not rampant communism—subverting from within the American values that were supposedly being defended on the hills of Korea."[12]

Yet even in the midst of this frenzy, signs appeared throughout the country that consistency would not stand in the way of a good deal. In Milwaukee, the largest city in Senator McCarthy's home state, the public committed itself financially to housing the Braves in County

Stadium. Baltimore, the largest city in the state that ousted Senator Millard Tydings—a McCarthy target—lured the Browns from St. Louis with the enticement of the use of a municipal stadium. Private enterprise was undoubtedly praised at numerous city functions in those communities; but if a dose of socialism was required to secure a major league franchise, then city officials were often willing to act like Marxists for the moment. Public housing in Los Angeles was proving less popular than public ballparks. Within a year of signing the city contract for Chavez Ravine, city council members bowed to the pressure applied by opponents of the housing program; and on the day after Christmas in 1952, the city council, in an 8–7 vote, attempted to abrogate its contract with the City Housing Authority.[13]

The Housing Authority, established to implement federal law, retained support from Mayor Bowron, labor unions, civil rights organizations, and other groups faithful to New Deal values. As in many other cities, advocating government aid to the disadvantaged had come to be associated less with Franklin Roosevelt than with the Kremlin. Even patrician supporters of American capitalism, such as Dean Acheson, were publicly ridiculed for being "soft on communism." When three members of the Housing Authority were questioned by a state senate committee on un-American activities about their membership in the Communist party and responded by invoking the Fifth Amendment, they were immediately dismissed from the agency, and the entire program was fatally compromised.[14]

The *Los Angeles Times,* far more powerful than television, framed its coverage of the issues in the tradition of Harrison Gray Otis. Not content simply to block the specific programs of the Housing Authority, the *Times* lobbied for nullification of the contract between the city and the federal government.

The *Times'*s advocacy reflected a confrontation between the values of democratic policy-making and the obligations of a contract. In a simple yet profound shift of opinion, the electorate, which once had seen public housing as an appropriate way to care for the less advantaged, had changed its mind. The reversal was complicated by the fact that in its initial enthusiasm the public had assumed a legal responsibility which, just because of a shift in the political winds, could not be lightly shrugged off, even though Congress had recently passed legislation permitting public housing contracts to be nullified by referendum.

In a portentous move, litigation was filed by housing supporters to determine if a contract could be nullified by a vote of the city council.[15] The council in turn called for a public referendum on the housing project. The vote was held on June 3, 1952, and a 60-percent majority rejected the contract. Voters apparently were innocent of or indifferent to a unanimous state supreme court decision six weeks earlier upholding the contract.

The California court had specifically rejected both the authority of the city council to abrogate the housing contract and the validity of a public referendum. Its opinion stated that "it was obviously never assumed, and certainly it was not authorized by law, that after the city council had declared the need, had given requisite approvals of a project under state and federal law, and had undertaken binding commitments, it would or could repudiate them."[16] The court construed the federal law authorizing referenda to defeat public housing contracts as having effect only in the future. The legislation, in the court's judgment, "does not contemplate a case where, as here, the city has approved the project and the federal agency has authorized the construction. It does not give the agency power to rescind prior authorization of construction or repudiate the obligations incurred thereunder."[17]

After Poulson defeated Bowron, he was legally compelled to execute the contract he had campaigned against, although local politics virtually precluded using the acreage for public housing. Poulson's solution was to purchase Chavez Ravine from the federal government, which sold the land to the city on the condition that it be used for some appropriate public purpose.[18] No apparent consideration was given to restoring title to the original occupants who through a decade of legal machinations continued to live in their homes, technically as squatters.

Roger Arneburgh, who was the Los Angeles City Attorney at the time, confirms that Poulson reached a political compromise with federal authorities under which some of the pending housing projects would be completed and others not.[19] Chavez Ravine was one of the latter. The city then took possession of the acreage with the provision that the land be put to some public purpose, although not necessarily housing. Years before Walter O'Malley even considered relocating the Dodgers, the team's future home was secured by powerful interests organized to block public housing.

In his memoirs, Mayor Poulson described his efforts to find some use for the land.

> For three years, I tried to get public groups interested in the area. I talked to sponsors of the opera house project, to zoo patrons, to horse show enthusiasts and others, but none was excited over Chavez Ravine. The only nibble I got was from a cemetery firm which wanted to buy the land for a burial ground.

Meanwhile, the place was inhabited by squatters and by a handful of small home owners whose goats, cows and chickens roamed about. Few others ventured into the hills, except evening spooners.[20]

No party was prepared to offer cash to convert the land into anything resembling "an appropriate public purpose."

This inability to find a public use may have been important legally. Arneburgh believes that the years of futility prior to the Dodgers' contract was a crucial legal point that distinguishes the interpretation of the housing law in Los Angeles from its interpretation in New York.[21] If Los Angeles had had to initiate condemnation proceedings for the purpose of selling the land to a baseball team, the courts might not have sustained such an interpretation. In fact, when the Dodgers took title to the Chavez Ravine property, the city dropped condemnation proceedings against twelve sites, forcing Walter O'Malley to purchase the plots on the private market.[22]

Grist for conspiracy theories surrounding the timing of Walter O'Malley's decision to move the Dodgers to Los Angeles can be found in O'Malley's first serious meeting with Los Angeles officials in October 1956. Having completed their final subway series with the Yankees, the team stopped in Los Angeles on the way to visiting Japan. During the layover, O'Malley met with Kenneth Hahn, the County Supervisor, to discuss the possibility of the Dodgers' moving to Los Angeles. Hahn recalls that he had been in New York during the 1956 World Series to see if the Washington Senators could be induced to move West.[23] This venture had been authorized by the County Board of Supervisors, which operated independently of city officials. The county anticipated a municipal stadium constructed with revenues drawn from county employees' pension funds. During Hahn's stay in New York, he received a note from Walter O'Malley indicating he would like to discuss a move to Los Angeles. Hahn reported that he shook hands with O'Malley on the transfer of the Dodgers during subsequent meetings in Los Angeles, although O'Malley cautioned the supervisor not to mention the decision publicly lest the outrage in Brooklyn be overwhelming. As subsequent events would prove, the agreement between O'Malley and Hahn could have been nothing more than an understanding in principle. Hahn himself says that once the Dodgers decided to move to Los Angeles, he withdrew from the picture, leaving the contract negotiations to the city.

In November 1956, as the Brooklyn Sports Center Authority continued to flounder, Poulson commissioned a study of Chavez Ravine by a group of civil engineers, who reported that the land might be used for the construction of a baseball stadium and recommended that $2 million be appropriated to that end in the mayor's 1957 budget.[24]

There were two major obstacles to this proposal. First, the unanswered question of whether a baseball stadium would conform to the requirement, still pertinent, that Chavez Ravine be put to an appropriate public purpose. Second, $2 million was only a fraction of the sum needed for a new stadium. For the city to finance the structure, additional revenues would have to be secured through bonds requiring two-thirds approval by the electorate.

Although Poulson played a pivotal role in bringing the Dodgers to Los Angeles, he was not blindly committed to bringing just any major league team to the city. Shortly after assuming office, he declined to make an offer to entice the Browns because, as he wrote, "They did not have a good team or baseball organization."[25] Several years later he also declined to pursue the Washington Senators, who were also looking for new quarters. Poulson's reluctance stemmed from his days as a congressman, when "I had seen them play too many times in Washington, D.C., and they were also 'tail-enders.' "[26] Poulson remained determined to secure only a top-caliber team.

Poulson was remarkably naive or else remarkably Machiavellian. He either simply assumed that a successful firm would willingly relocate in an untried market, or he foresaw the difficulties facing the Dodgers in New York and positioned Los Angeles to take advantage of them. Judging from future negotiations between Poulson and O'Malley, naivete is the more plausible explanation. But there was a rational basis for Poulson's optimism. He was aware that the first modern franchise shift in professional sports had occurred in 1946 when the Cleveland Rams of the National Football League relocated in Los Angeles. Within five years they won a championship, and Poulson concluded that as far as baseball was concerned, "Los Angeles rated the best because we had championship teams in football and track and other athletic activities."[27] Poulson later recalled sitting in the bleachers and rooting for the Dodgers during the 1955 World Series, but he dismissed any speculation that he had designs on the team at that time.

The recommendation to construct a baseball stadium in Chavez Ravine was made at a time when major league ball had recently spread to Baltimore, Milwaukee, and Kansas City. In a short time, Los Angeles would become the focus of national attention as a prospective major league site. The attention, of course, would center on whether the Dodgers would transfer to the West Coast, and in the heat of the controversy many would forget the rich history of baseball in Los Angeles and of an intriguing earlier opportunity for bringing major league ball to the West.

Organized baseball was played in California as early as 1845.[28] Leagues were established by the 1880s, and a Los Angeles entry won its first championship in 1892, just a year after Brooklyn had secured its first pennant. The Los Angeles franchise was among the most successful in what became the Pacific Coast League, winning eleven pennants from 1887 through 1938. Southern California became such a resource for professional clubs that a second team was established in Hollywood. The prosperity of baseball in the West occurred somewhat in isolation from the rest of the country, but many of the game's greatest players, including Tris Speaker and Joe DiMaggio, started their careers with minor league teams in California, which remained a rich resource for the major leagues until the technologies of communications and transportation shrank the country and thereby reduced the impediments of great distance and imposing mountain ranges.

By 1940 advances in radio and rail travel made the development of major league ball in California feasible. To secure an established major league team, a city might look to one of the struggling franchises. In that regard, no team struggled with greater futility than the St. Louis Browns. Between 1902 and 1940, the Browns won no pennants and had, in fact, only nine winning seasons. By the end of the 1930s they competed regularly with Philadelphia for the American League cellar. The Cardinals, in contrast, had become the dominant team of the National League. Thus St. Louis was one of the first of the two-team cities to face an imminent franchise shift. After the 1941 season, the Browns were prepared to transfer their version of major league baseball to Los Angeles.[29] The move was abandoned after the United States entered World War II, since it was not certain initially if baseball would even be played during the conflict. The war brought unprecedented prosperity to the Browns and in 1942 they finished third in the American League. They slipped to sixth place in 1943, but in 1944 they won their only pennant and provided St. Louis with its only crosstown World Series as the Cardinals also captured a league championship. The Browns reverted to form, losing the Series in six games, dropping the final three contests.

Baseball during the war years has been characterized as the "Spartan Seasons."[30] Not only were many star players pressed into military service, but millions of fans were in uniform and unable to provide financial support for their teams. The end of the war created an expectation that booming economic times would bring unprecedented prosperity to the game. Notice was served almost immediately that the existing structure of the major leagues was insufficient to accommodate the anticipated changes. The Pasquel brothers of Mexico enticed several major league players to challenge the reserve clause and leave their teams to play in Mexico for significantly higher pay.[31] In 1946 the players wrung a Bill of Rights from the owners that established certain fundamental benefits, their independence was demonstrated further when the Pittsburgh Pirates threatened to strike and by the formation of the Major League Players Association which, in recent years, has been such an important force in the game.

Challenges to the status quo were not limited to the labor market. In December of 1945, the Pacific Coast League announced its intention to operate as a third major league for the 1946 season.[32] The owners of Coast League teams did not intend a rebellion; they petitioned their fellow minor leagues for approval, then sought the blessing of the National and American leagues during their annual winter meetings. On December 12, the major leagues rejected the request, alleging it would be unfair to the players. Clarence (Pants) Rowland, the Pacific Coast league president, replied that the majors were merely postponing the inevitable. The following year the Coast League reintroduced its proposal, and the majors agreed to study the matter.[33]

Several spokesmen in Los Angeles, including County Supervisor Leonard Roach and *Los Angeles Examiner* columnist Vincent Flaherty, revived efforts to attract a major league franchise. What was known as "the St. Louis problem" remained to be solved, and the prospect of the Browns' relocation to Los Angeles was being reconsidered, now that the war had ended. These efforts conflicted with the Coast League's desire to achieve major league status for all its members, including those in smaller markets.

The Pacific Coast League included four teams in the Los Angeles and San Francisco areas and another four in San Diego, Sacramento, Portland, and Seattle. The loss of markets in Southern California and the Bay Area would threaten the integrity of the entire league. Coast League officials were determined that the major leagues would come to the West Coast through the upgrading of the entire league's status rather than through the incursion of an existing major league franchise to a single city. The possibility of the Browns' move to Los Angeles was a serious threat. The Pacific Coast League in the 1940s occupied a lofty place in the hierarchy of professional baseball, holding the highest rank in minor league ball, the Triple A classification. Only two other such leagues existed, and they included a total of twenty-four teams. Subordinate classifications included Double A with three leagues and twenty-two teams, and Single A with two leagues and fourteen teams. These same classifications exist today, but thirty years ago minor league baseball was far more elaborately constructed than is currently the case. Players who could not qualify for the more exalted leagues could perhaps catch on with a team in one of the leagues with a Class B, C, or D ranking.

We saw earlier the controversy between Judge Landis and Branch Rickey about whether the development of farm systems would cripple or sustain the minor league franchises. The record since 1960 reflects a remarkable contraction in the number of minor league teams, as their purpose has narrowed. Once they had provided the only exposure local communities had to the national pastime; by now that function has been taken over by television as well as by the expansion of the major leagues. Minor league teams exist today almost exclusively to serve the interests of major league franchises. They consume significant resources provided by their major league patrons, and consequently employ only those players with major league potential.

At the 1946 winter meetings, when the major leagues effectively tabled the Pacific Coast League's bid to upgrade its status, they also acceded to a demand that any major league team that invaded a Coast League city must pay an indemnity to the local minor league franchise as well as to the rest of the league.[34] This was a formidable impediment for the weaker teams in the majors, and it helped stymie efforts to reconstitute the transfer of the St. Louis Browns. Buoyed by the partial success of the indemnity requirement, the Coast League pressed its case for higher status. In August 1947 the league petitioned for exemption from the major league draft as a critical step in its attainment of major league stature.[35] This move reopened the controversy that had raged between Rickey and Landis over the purpose and interests of the minor leagues.

The first draft agreement permitting major league clubs to call a player up from a minor league franchise was established on March 1, 1892, as part of a general agreement among the various leagues of professional baseball.[36] The terms of the draft were periodically renegotiated between the major and minor leagues with the final agreements binding for all parties. In 1919 the minor leagues abrogated the draft provisions, and when another general agreement was reestablished in 1921, participation in the draft was left voluntary. The Pacific Coast League was one of five leagues to abstain from the draft, but by 1930 the major leagues had pressured the minors into accepting a universal draft by threatening to withdraw all ties to the five leagues.

The chief concern of the minor leagues was, of course, financial. They complained that after spending money to develop a player he could be called up before the original team had recovered its investment. A new agreement in 1931 protected minor league teams by making their players ineligible for the draft until they had played a certain number of years, four seasons in the case of the highest-rated minor league franchises. In 1947 the Coast League argued that it merited exemption from this agreement because it drew a larger attendance than other Triple A leagues, paid higher salaries, and wanted to pursue its destiny as a future major league.[37] Other minor leagues objected on the grounds that Coast League attendance figures were misleading, due to a significantly longer playing season (186 games as opposed to 154) and that their attendance figures included women and children, who could gain admittance to the ballpark by paying only the tax on the regular ticket fee.[38] The Coast League replied that no other minor leagues could be taken seriously since they were merely farm teams of the major league clubs. In the Coast League only the Los Angeles Angels were owned by a major league team (the Chicago Cubs), while the Hollywood Stars retained a working agreement with the Pittsburgh Pirates. The other teams were all independently owned and run in accordance with Kenesaw Landis's prescription for minor league baseball.

After the 1947 season, the Pacific Coast League modified its position, calling for retention of the draft but with the protection of players extended from four to six years and payment for each player raised from $10,000 to $25,000, the elevation of the PCL to a status higher than Triple A, and the acceptance of the Coast League as a major league in three years.[39] In the winter meetings, the other minor leagues referred the decision to the major leagues. Although the necessary three-fourths majority of the minor leagues failed to support the bid, a plurality did approve the plan.[40]

The majors not only turned down the petition; they also showed support for a new proposal by Commissioner Happy Chandler who, as promised, spent the year considering the Coast League's grievance. Chandler argued that the appropriate way to bring major league ball to the West Coast was by expanding the number of American and National league teams to ten each. The additional clubs could be established in Los Angeles, Hollywood, San Francisco, and Oakland. The National League approved the plan unanimously, but the American League rejected it 5–2, with Cleveland abstaining and the New York Yankees voting in the minority.[41]

The strong support for Chandler's proposal indicated that the majors were prepared to skim the cream from the Pacific Coast League in order to control expansion through the oligopoly of the National and

American leagues. The majors then completed their rebuff of the Coast League by rejecting its proposed draft modifications. This decisive action curtailed for several years the efforts of the Coast League to upgrade its status. At the outset of the 1950 season, Paul Fagan, owner of the San Francisco Seals, suggested that the Pacific Coast League might withdraw from organized baseball and operate independently if its concerns were not satisfactorily addressed by the major leagues.[42] His warning led to meetings with the commissioner, who repeated the platitudes of *Sporting News* editorials regarding the legitimate interests of the Coast League. At the end of the 1950 season, the major leagues announced that they might confer "Four A" status on the Coast League, elevating it to a unique position in the game's structure.[43]

This ploy succeeded for a year. In August 1951, the Coast League again called for the elimination of the draft but added that if such action were not taken by the end of the year, the league would indeed withdraw from organized baseball.[44] The threat assumed significant credibility because it came at the same time the Judiciary Committee of the House of Representatives was holding hearings on baseball's exemption from antitrust law, focusing specifically on the status of the Pacific Coast League. Committee members were also curious to know why, in spite of the tremendous demographic changes in the twentieth century, major league ball continued to be limited to a few Northern and Midwestern cities.

Even P. K. Wrigley, owner of the Chicago Cubs and the Los Angeles Angels, agreed it was time to end the monopoly enjoyed by the major league franchises.[45] He later modified his blunt remarks, but his confreres, under a new commissioner, Ford Frick, were ready to cut their losses rather than risk any legislative tampering with the antitrust exemption. During the 1951 winter meetings, Frick announced a scheme to upgrade minor leagues to major league status.[46] A new classification above Triple A was established for leagues or groups of at least eight teams whose aggregate markets included at least 10 million people, whose ballparks had an aggregate capacity of 120,000 and who had an average paid attendance of over 2.25 million for the preceding five years.

The only Triple A league to qualify for this "open" classification was the Pacific Coast League, but stringent barriers were imposed on its further elevation. Frick announced that leagues in the open classification would be considered for major league status only if they met the following conditions: an aggregate market of 15 million, ballpark capacity of more than 25,000 for each franchise, and paid attendance in excess of 3.5 million for each of the previous three years. The Coast League met the first two qualifications, but failed the third; moreover, attendance had declined precipitously the previous two years. Attendance figures for the years 1947–51 indicated the following: 1947—4,068,432; 1948—3,661,166; 1949—3,751,929; 1950—3,172,718; 1951—2,282,030.[47]

Frick's third barrier raises questions about the sincerity of his proposal. Not only was Coast League attendance falling drastically below the required mark, but the decline was at least in part traceable to controls exercised by the major league clubs, which cut into traditional minor league markets with a flood of televised games. Frick tossed a sop to the Coast League by extending draft protection to five years for clubs in the open classification. This new arrangement ended the six-year effort by the Pacific Coast League to bring major league baseball to the West through the upgrading of teams with historic ties to Coast cities. Coast League attendance continued to decline during the 1950s and thus eliminated any basis for renewed efforts to achieve major league status.

One can easily imagine how baseball might have expanded nationally after World War II by adding new leagues rather than by extending the old order. The Pacific Coast League was an obvious possibility as a third major league, and the eventual emergence of a league in the South would have made the game truly national. Such an arrangement would have preserved historic ties and rivalries in place of the sometimes haphazard expansion that has, in fact, ensued. Legitimate questions were raised during the 1940s about the adequacy of the Pacific Coast League's markets, the size of the ballparks, and the quality of play; but the major league owners were more seriously disturbed by the prospect of surrendering control of an industry that had been in their possession more than forty years. That threat having been blocked, the expansion of major league baseball to Los Angeles would come to depend on the interests of the established major league owners.

In the winter of 1956–57 the Dodgers were rebuffed by the New York City Board of Estimate in their attempt to appropriate funds for a study of the Atlantic-Flatbush site, while city and county officials in Los Angeles apparently offered the Dodger president everything but a toll-booth on the Harbor Freeway. Whatever any of the participants thought they had decided during those months, the Dodgers' fate was far from settled. Los Angeles had to clear many obstacles before it could offer a legitimate proposal, and New York still had a few cards to play before finally rejecting O'Malley's designs. Winter had been a time when wishful thinking overcame sound judgment on the part of Los Angeles officials and O'Malley himself. Spring resumed the process of hard bargaining and difficult choices. To Roger Arneburgh, the city attorney, fell the task of telling Walter O'Malley that the enthusiasm he had received from the mayor and county officials could not be sustained under the law. Arneburgh recalls that O'Malley had been offered five hundred acres of land in Chavez Ravine and also tax exemptions.[48] At a meeting in May 1957 in Los Angeles O'Malley learned that such an offer was illegal and improper, and that the entire proposal would have to be renegotiated.

In February 1957 Walter O'Malley purchased from the Cubs their minor league franchise, the Los Angeles Angels, and the Angels' home park, Wrigley Field, near downtown Los Angeles.[49] Just as he had forced the action in New York with his announcement in August 1955 that the Dodgers would schedule games in Jersey City, now O'Malley inspired city officials in Los Angeles to formulate a plan that would lure him across the country to a market fifteen hundred miles from the nearest major league franchise.

O'Malley's acquisition of the Los Angeles minor league team raised questions about his motives in his negotiations. At federal antitrust hearings in the spring of 1957, O'Malley was asked why he had purchased the Los Angeles Angels. He replied, "I bought the Los Angeles club when I have every sound reason to believe that there would be no solution in Brooklyn."[50] Later in the hearings, O'Malley insisted he retained some hope that the Brooklyn stadium could indeed be built; but mistakably the meager sum appropriated in December 1956 by the Board of Estimate for further studies of the Atlantic-Flatbush site helped inspire O'Malley to purchase an alternative home for the Dodgers. What might be called "the O'Malley Devil Theory," advanced by Harold Parrott and embraced by Peter Golenbock, purports that the Dodger president suckered the Cubs' Philip Wrigley into abandoning the property in Los Angeles. "All O'Malley did was con Wrigley into taking, in a straight swap—lock, stock, and ballpark—the Fort Worth franchise in the Texas League."[51] In Parrott's view the exchange was hardly equitable; Wrigley was annoyed with the city of Los Angeles, and O'Malley happened to run into him at a propitious moment.

Another interpretation emphasizes the role of television in baseball during the 1950s. This theory was stated by John Old of the Los Angeles *Herald-Express*, who contended that Wrigley cooperated with O'Malley in an effort to break up the multiteam cities that had characterized the distribution of franchises from 1900 into the 1950s.[52] Single-team cities would make optimum use of television by broadcasting all road games while blacking out home games. With two-team cities, conflicts would arise if one team televised road games on the same nights

that the other club tried to attract fans to its home games. The solution
was to have only one team in each major city. In this complex arrangement, Wrigley was not the victim of O'Malley's scheming but a fellow
conspirator who aided in removing the Dodgers and Giants from New
York in exchange for help in forcing the White Sox to a new location
in Minneapolis (territory owned by the Giants) or Fort Worth (formerly
a Brooklyn, now a Chicago Cub, site).

In any case, twenty years after the Dodgers and Giants left New
York, Chicago is still a two-team city, and the Dodgers and Giants both
agreed to American League franchises in Los Angeles and the San
Francisco Bay area. Minneapolis later became the home of the old
Washington Senators, and Fort Worth received the expansion Washington Senators, now the Texas Rangers. Old's and Parrott's interpretations are not necessarily wrong, but history has not borne them out.
And in light of the feeble support for a new stadium in Brooklyn, one
can reasonably take O'Malley at his word: he acquired the Cubs' rights
to the Los Angeles territory because he had determined that an accommodation with New York officials was unlikely. Considering that the
agreement between O'Malley and Wrigley was concluded just a month
after the meager appropriation by the Board of Estimate, compelling
evidence supports O'Malley's view of his New York prospects.

At the same time O'Malley acquired the Los Angeles Angels, he began to negotiate with city and county officials in Los Angeles. He met
with Mayor Poulson and other officials at Vero Beach during spring
training.[53] Poulson described the session as "a sparring match," although from the outset the participants overestimated each other's
strength.[54] The Dodgers had already sold Ebbets Field and their lease
was due to expire in 1959. Because they had been rebuffed by the New
York Board of Estimate, their only sure option for the immediate future
was scheduling home games in Jersey City. Somehow, despite that
predicament, Los Angeles officials concluded that O'Malley was holding all the cards. Poulson later recalled that "One of our officials promised O'Malley the moon, and Walter asked for more. You couldn't blame
him. He had a valuable package in the Dodgers and he knew it. I assured him we wanted desperately to get the team, but made it clear
we would have to come up with a plan that wouldn't get all of us run
out of the city."[55] In the course of these meetings, Poulson mentioned
Chavez Ravine as a possible site for a new stadium. O'Malley was already familiar with the area from his earlier meeting with Hahn and
had sent an engineering firm to study the site during the latter part of
1956. The Dodgers' president, himself an engineer, was enticed by the

possibilities of so much centrally located land near a confluence of free-ways.

Perhaps as a test of Mayor Poulson's eagerness, O'Malley allegedly changed a fundamental objective of his previous stadium pursuits. "O'Malley," Poulson remembered, "gave us no promises at Vero Beach, but did say that if he were to consider our city at all, he would expect us to build a ball park. Land didn't interest him in the least."[56] This position is quite contrary to the record of negotiations in New York for a replacement for Ebbets Field. There, O'Malley had sought the con-demnation of land for his subsequent purchase and for private con-struction of a stadium. Only after Robert Moses rejected this bid did O'Malley settle for the prospect of being a tenant in a facility built and owned by the Brooklyn Sports Center Authority. In his first serious negotiations toward moving the Dodgers out of New York entirely, he proposed the construction of a municipal stadium.

Poulson noted that O'Malley subsequently denied that he had ever raised this point, then added, "All I can say is that he was a very smooth talker and left me with the impression that he would want the City to build a stadium."[57] After the Vero Beach meetings, Poulson intended to see if the city could comply with what he at least believed was an important demand from O'Malley.

In April 1957, O'Malley visited Los Angeles, ostensibly to inspect his new property, Wrigley Field, where the Angels played their Pacific Coast League games.[58] Poulson records that O'Malley reiterated a desire for a municipal stadium, suggesting a "Milwaukee formula" whereby in return for playing in a publicly owned stadium, the city could tax re-ceipts and concessions. Poulson ordered a survey of Chavez Ravine to determine the cost of a stadium. In a curious bit of municipal finance, the city borrowed $5,000 from Howard Hughes to help fund the engi-neering study. The billionaire was never repaid.[59]

The report concluded that a stadium would cost anywhere from $10 to $15 million, a prohibitive amount, since taxes gained under a "Mil-waukee formula" would be inadequate to repay funds borrowed from the County Workers' Pension funds, the potential source of construc-tion revenues. An alternative funding source, municipal bonds, was unfeasible because the city had a history of rejecting such proposals. As Poulson wrote in his unmistakable style, "For less money than what the ball park would have cost, voters beat airport bonds twice before I came into office. They even licked sewers. And Los Angeles could have used them."[60]

By the spring of 1957, Los Angeles knew it had a chance to acquire not only a major league franchise but one of the most attractive in the game. Newspapers, business, and civic groups applied pressure on Poulson and others to secure the deal. In May, Poulson visited O'Malley in Brooklyn and brought him up to date on West Coast developments. The mayor explained that the city couldn't build the Dodgers a stadium, but it could provide the land at Chavez Ravine for a meager sum, and the Dodgers could construct their own park.

Presented with what he had long sought in New York, O'Malley, according to Poulson, "hit the ceiling. 'I already have one ball park there!' he exclaimed. 'What am I going to do with two?' "[61]

O'Malley pointed out that no one had built a park with private funds in more than thirty years, then mentioned a constraint that may elucidate his puzzling behavior. A move to Los Angeles, O'Malley informed Poulson, required that an indemnity be paid to the Pacific Coast League, a consideration that obviously wouldn't apply if the Dodgers could work out an agreement with New York officials. The record is clear that in New York O'Malley favored a private stadium on land that he would purchase with the city's assistance. The evidence on this point in Los Angeles is mixed.

Dick Walsh, currently the general manager of the Los Angeles Convention Center, was employed by the Dodgers in 1947 after service in the Navy and later became O'Malley's liaison to many public officials in Los Angeles, Walsh's hometown. He recalls O'Malley giving him explicit instructions to see if a municipal stadium could be constructed for the Dodgers and offers the explanation that O'Malley realized that moving the team involved enormous risks which he wanted to minimize as much as possible. He would be less tied to Los Angeles if the Dodgers played in a public stadium.[62] Kenneth Hahn, on the other hand, says that when he met with O'Malley in October of 1956 he specifically offered the Dodgers a municipal stadium, and that the Dodger president rejected the proposal.[63] Hahn maintains that O'Malley insisted on building his own facility. O'Malley's background as an engineer figures importantly in this explanation, which holds that he saw the stadium as an opportunity to make a permanent impact on the game. Red Patterson, a close aide to O'Malley at the time of the move, remembers nothing to indicate the Dodgers might play in a municipal stadium.[64] Patterson asserts that O'Malley long expressed a desire to build his own stadium. If everyone's memory is accurate, we conclude that O'Malley simply told different things to different people.

Since O'Malley had registered objections to playing in a public stadium for several years, the stance which he took with Poulson suggests he was posturing for negotiating purposes. In reality, O'Malley seems to have been eager to own his own stadium because of the control it would afford. Perhaps he still anticipated the growth of pay television systems and feared the interference of the state in what might prove a highly lucrative venture. The Dodger president was never captivated by short-term profit when long-term fortune seemed assured. After O'Malley's outburst, Poulson again underestimated the strength of his own position. In his memoirs, the mayor wrote: "Since O'Malley needed us much less than we needed him, he obviously held the trump cards. In the course of our discussion, he asked if maybe he could sell Wrigley Field to the city. I knew in my heart that the only plausible solution was a trade of Wrigley Field for Chavez Ravine, but I also knew that in dealing loosely with city property I was gettin [sic] in over my head and was playing with political dynamite. I told Walter I would do the best I could."[65]

In case Poulson failed to recognize his own limitations, they were pointed out to him upon his return to Los Angeles when he met with the publishers of the *Los Angeles Times* and the *Examiner*, who "concluded that I was no business genius, because they suggested that I get as my representative some top-notch negotiator who understood real estate values and who could hold his own in a bargaining match with the wily O'Malley."[66]

Poulson called on Harold McClellan, a business leader and former assistant secretary of commerce. McClellan met with City Attorney Arneburgh and Samuel Leask, the Chief City Administrator, to devise a strategy for negotiating with O'Malley. Poulson suggested the possibility of exchanging Wrigley Field for Chavez Ravine, and they integrated that prospect into the core of the final deal. McClellan wouldn't be appointed to his post as negotiator until late July, and he indicated he needed time to master the pertinent facts and formulate an appropriate strategy for the negotiations. Nonetheless, he was a skilled executive and served without pay during months of difficult bargaining. On August 26, McClellan reported to the Los Angeles City Council that the city stood a good chance of drawing the Dodgers if it could put together expeditiously a reasonable package.[67] O'Malley was still negotiating with New York officials, who had been chastened a week earlier by the announcement that the Giants were leaving New York for San Francisco.[68]

One of the obstacles for Los Angeles was that O'Malley wanted three hundred and fifty acres at Chavez Ravine; the city held a considerably smaller parcel. Some city council members bridled at O'Malley's request—they had understood the Dodgers to be after only two hundred acres. McClellan replied that O'Malley had originally asked for six hundred acres in the Vero Beach meeting of March.[69]

Supporters of the Dodgers' move emphasized the need for swift actions, and stressed the economic returns the city would reap by putting Chavez Ravine on the tax roles for the first time and by providing a stimulus to local businesses. The time constraint was indeed severe. National League rules required teams to declare a franchise move for the following season no later than October 1. The Dodgers had received permission to move during the league's May meeting, but were still obligated to declare their intentions by the October deadline. O'Malley, of course, needed a binding commitment from Los Angeles before he could cut his ties to New York. Los Angeles was faced with providing such a commitment within a month after McClellan made his report to the city council.

Paul Zimmerman, sports editor of the *Los Angeles Times*, expressed his paper's sentiments in a September 12 column that argued, "Time is running out on the chances of the city of the Angels getting major league baseball by next year."[70] Zimmerman allowed that legal complications stood in the way, but he pointed out that San Francisco had secured the Giants with apparently less difficulty than Los Angeles was having. He also expressed concern that since the Giants had announced their move New York would now redouble its efforts to keep the Dodgers.

In fact the obstacles that faced the Dodgers following the Giants' move had less to do with the details of an agreement than with the ratification process involving the Dodgers and Los Angeles. All parties understood that if the Dodgers were to come to Los Angeles they would receive approximately three hundred acres of land at Chavez Ravine for which the city would commit $2 million for grading and the county another $2.7 million to construct access roads. In return the Dodgers would give the city Wrigley Field, valued at over $2.25 million, develop and maintain for twenty years a forty-acre public recreation area at Chavez Ravine, and pay an estimated annual $350,000 in property taxes. The city and the Dodgers agreed to share oil and mineral rights, although no such resources had been found at Chavez Ravine. The deal was acceptable to O'Malley, who faced the October 1 deadline with no official commitment from Los Angeles, nor a feasible alternative offered by New York.

At this juncture the critical factor was a government obstacle we have met before, the requirement of an extraordinary legislative majority to guard against impassioned and unwise democratic policy making. The

city council had to approve the agreement with a two-thirds majority—
ten votes out of fifteen. For some time a clear plurality had favored the
agreement but no one could be sure about the precise strength of the
opposition. The struggle for the crucial votes was intense, as Mayor
Poulson recalled: "I was sure of eight. Four were against, one was out
of town, and two were on the fence."[71] Council members John Holland
and Pat McGee led the opposition, objecting that the city was giving
away too much. Unlike most city council members who thought of the
electorate as baseball fans, Holland, McGee and their allies considered
the voters primarily as taxpayers who would object to a boondoggle.[72]

In Poulson's view, Pat McGee was motivated less by his concern for
the public weal than by the interests of a small group in San Diego tied
to the Pacific Coast League.[73] This group was headed by Arnholt Smith,
owner of the San Diego Padres of the Coast League and later the major
league franchise of the same name. McGee and Holland adopted the
tactic of not directly opposing the Dodgers' shift to Los Angeles but of
calling for a better deal for the city. Attracting the Dodgers had become
a popular issue for city residents, but the prospect of saving money
retained its age-old appeal.

On September 16 the city council voted 11–3 to approve a resolution
authorizing McClellan to submit for the Dodgers' consideration a plan
in which "about 300 acres, more or less" would be granted to the franchise in Chavez Ravine in exchange for Wrigley Field.[74] The plan also
considered mineral rights, recreational facilities, and expenditures for
improvements at Chavez Ravine. The 11–3 vote was somewhat misleading, since Councilman Holland voted in the majority in the vain
hope he would be able to reopen debate on the proposal the next day.
Moreover, the resolution itself fell short of the clear signal O'Malley
needed, since it merely authorized the submission of a plan to the
Dodgers rather than a final commitment to an agreement.

These legislative machinations were precisely what the opponents
relied on as they sought to block the Dodgers' move. By offering
amendments to the plan, they could conceivably force a renegotiation
of the entire agreement. Another tactic was a move by Councilman
Holland to protest before the entire city council that the grading of land
at Chavez Ravine would endanger the nearby Los Angeles Police
Academy. His move was sidetracked in a three-member subcommittee
controlled by O'Malley supporters.

On September 25, Los Angeles Times reporter Frank Finch wrote from
New York that O'Malley was beseiged "on all sides" as the National
League deadline approached.[75] Finch pointed out that any extension of
the October 1 requirement for an announcement of the move would

have to pass unanimously, and that "several owners are known to be strongly opposed to their league deserting the rich New York territory." Finch also noted that Commissioner Frick and some National League officials "have expressed puzzlement at the dillydallying in Los Angeles." The article continued in an editorializing vein: "They are wondering tonight if the City of the Angels really wants to be a big league town in the senior circuit. And so are a lot other people, we might add."[76]

One reason for the *Times*'s consternation was a partial victory by the opponents of the move. Holland succeeded in gaining consideration of a motion for an appraisal of Wrigley Field and Chavez Ravine wholly independent of the estimates already prepared by the city's Board of Public Works. Councilman James Corman charged that Holland's motion could take months to effect and was merely a ruse to block the entire offer. Holland furiously denied the charge and demanded an apology.[77] The *Times*'s enthusiasm for the Dodgers continued to influence the objectivity of its reporting as Paul Zimmerman wrote that the council's consideration was caused by Holland's "continued attempt to thwart the best major league deal any city has been offered in 20 years."[78]

The Dodgers' supporters were frustrated because although they appeared to have the necessary votes to meet even the stringent two-thirds requirement, another parliamentary barrier now threatened to block passage of the plan. The city charter required that the type of resolution needed to secure the deal with the Dodgers could pass on its first reading only if approval was unanimous; otherwise seven days must go by before a second reading was permitted and the two-thirds vote would be sufficient. A first reading of the measure was scheduled for Monday, September 30, the day before the Dodgers were required to announce their intention to move to Los Angeles or stay in Brooklyn. A unanimous vote was out of the question, and opponents of the plan intended to offer a series of amendments, any one of which could derail the entire negotiations. Mayor Poulson promised to achieve the ratification and secure the Dodgers for Los Angeles.

The high theater of politics peaked when Gordon Hahn, a councilman and brother of the county supervisor, returned from a vacation to announce he had a plan sure to save the day.[79] Hahn proposed that if ten members voted for the plan on its first reading the city council would consider itself morally bound to provide that same margin on the subsequent vote a week hence. Hahn also told the press that his staff had informed him that prior to his return the vote was 9–5, and that he represented the crucial tenth vote. Hahn's ploy to bind the members of the city council failed to carry, but the vote still was considered a harbinger of what would occur on the second reading seven days later. The *Times* and Los Angeles city officials all viewed the first vote as symbolic because of the unachievable requirement for unanimity. They assumed that the Dodgers would be content with a two-thirds majority, the margin sufficient for the second reading.

The Sunday before the vote, the *Times* urged adoption of the resolution in its lead editorial.[80] On Monday, Paul Zimmerman reviewed the success of professional football and horse racing in Los Angeles, and concluded: "The question before the Council today is whether it is big league. Everything about sport in Los Angeles is."[81] Deliberations before the city council began on the morning of September 30 and lasted through the evening; even a bomb scare failed to quell the debate. Supporters faced a midnight deadline in extending the assurance to Walter O'Malley that the city would officially adopt the resolution.[82] In his memoirs Poulson recalled the tension of the extraordinary meeting: "I was up against it. On September 30, the Council was still arguing over the contract and not yet ready to vote. My leaders in the Council were Roz Wyman and John Gibson, who carried the discussion late into the night. All the while, I was seated in my office in another section of the City Hall, listening to the arguments over the intercom. Uneasily, I kept looking at my watch. The debate went on, and it was obvious that the vote would be delayed again. It was now 11:50 p.m."[83]

At this point, Poulson took bold action. In the presence of reporters, he telegraphed Warren Giles, president of the National League, to announce that he "had mustered the necessary ten votes." As the mayor confesses in his memoirs, "Actually I hadn't, but I couldn't gamble on blowing the whole deal."[84] On October 1, the *Times* dutifully reported the mayor's announcement along with the angry reaction of McGee and Holland when the telegram to Giles was read to the council by Councilwoman Wyman.[85] The only immediate effect of the telegram was to forge a consensus in the council in favor of adjournment. Poulson's maneuver, however, provoked an uproar, John Holland calling it "a breach of etiquette."[86] Even the mayor admitted, "My opponents were suspicious—with good reason."[87] Whatever the propriety of Poulson's conduct, it certainly stands in contrast to the activities of New York officials, none of whom was willing or able to force a decision to keep the Dodgers in Brooklyn. Moreover, Poulson demonstrated that even though the mayor's office in Los Angeles has been characterized as relatively powerless, it could effect change through skilled manipulation of public opinion.

For all the mayor's efforts, O'Malley was not idly waiting for the council's action. At the critical October 1 meeting at which O'Malley was to announce his decision to move or stay, National League owners voted to provide an additional two weeks for the Dodgers to complete an agreement with either New York or Los Angeles.[88] The second reading of the resolution was scheduled for Monday, October 7. To preclude any final snags, the Dodgers sent a club official, Henry J. Walsh, to answer any questions posed by council members. Walsh attempted no public persuasion of the uncommitted, asserting only that decisive action should be taken as soon as possible.[89] The October 7 meeting was another day-long affair. Rhetorical bombast competed with impassioned pleas, mostly in opposition to the resolution.

In the midst of this, Poulson prevailed on Wyman to call O'Malley for a definite commitment to bring the Dodgers West if the contract was approved.[90] Against her better judgment, Wyman placed the call and found that O'Malley was full of misgivings. Not only did the Dodgers' president refuse to make the commitment, he told her that he preferred to stay in New York. He stressed to her the risks involved in such a relocation, including the poor attendance for PCL games in Los Angeles. To Wyman's dismay O'Malley added that, having lived in New York all his life, he felt strong attachments to the city and found them difficult to break. This telephone conversation can, of course, be dismissed as mere posturing designed to influence the council vote. Wyman still believes, however, that O'Malley was sincere. In any case, she declined to discuss the conversation with her colleagues for fear it would turn them against the contract.

The argument that contends O'Malley betrayed Brooklyn ironically supports in part Wyman's interpretation of her talk with O'Malley. The fact that the most financially successful team in baseball would willingly move to such an uncertain market would give understandable pause to anyone contemplating leaving a setting as lucrative as Brooklyn. O'Malley may well have been a shrewd and skilled negotiator, but his skills were inadequate in New York, and second thoughts about the entire venture seem entirely appropriate.

Finally the resolution passed, 10–4, and the offer to bring the Dodgers to Los Angeles was official.[91] The months of debate had reached such acrimonious levels that Wyman had begun to receive death threats.[92] One might have expected that the passage of the resolution would end the controversy and restore calm to the city, but the council vote was just the beginning.

In a common version of the story, when Robert Moses and Mayor Wagner refused to cave in to Walter O'Malley's exhorbitant demands, the Dodger president suckered gullible Californians who, in their eagerness to attract the team, gave away the store. This rendition has the symmetry and passion of a morality play but, as we have seen, the record is considerably more complicated.

By February 1956, New York had had two eminent prospects for securing the Dodgers' future in Brooklyn. The first was Robert Moses's option as chairman of the mayor's Slum Clearance Committee to invoke his powers under the Federal Housing Act to assemble at Atlantic and Flatbush the parcel of land that O'Malley could have then purchased for the construction of a stadium. The second opportunity was to erect a public stadium under the auspices of the Brooklyn Sports Center Authority. Since this course required the cooperation of the mayor, the city council, and the Board of Estimate, as well as the governor and the state legislature, it was necessarily a far more tortuous path. O'Malley's interests were blocked completely on the first prospect, and were slowly expiring on the second. For the Dodger president to retain control of the club's future, new options needed to be created; accordingly, O'Malley opened a tie to Los Angeles through the purchase of the Angels.

In California, meanwhile, long-standing interest in acquiring a major league franchise had combined with a keen desire to do something about the white elephant of Chavez Ravine. City engineers had concluded even before O'Malley purchased the Angels that the area could serve as a site for a baseball stadium. The evidence further shows that Chavez Ravine had not been acquired for Walter O'Malley's benefit; rather, federal law required the city to reserve the acreage for an appropriate public purpose after the original public housing project had lost support. A number of observers wrote during the winter of 1956 that the Dodgers might eventually relocate in California, but these were mere conjectures that associated a team in search of a new playing field with a market which had been seriously interested in major league baseball since before World War II.

The interpretation which argues that Walter O'Malley orchestrated events in Los Angeles by manipulating the officials of that city ignores the highly pertinent histories both of baseball and the public housing controversy in Southern California. The overemphasis on O'Malley's role misses the vital contribution of major political actors in Los Angeles to the transfer of the Dodgers. The city council vote of October 7, 1957, culminated a concerted effort by city leaders to overcome formidable opposition to the final contract. Poulson, Wyman, McClellan, and others faced legal, political, and economic constraints similar or identical to those confronting New York officials. In Los Angeles those obstacles were overcome not because they were less imposing, nor even because the city's officials were uniquely skilled; rather, Poulson and the other advocates were committed to attracting the Dodgers, while their counterparts in New York were indifferent or hostile to the measures necessary for keeping the team.

When New York realized that the threat from Los Angeles was real, officials renewed efforts to keep the Dodgers in Brooklyn, or at least in a nearby community. Until those efforts were fully spent, Los Angeles could not be sure that its arduously constructed proposal to Walter O'Malley would finally succeed.

6

New York: Endgame in Brooklyn

The announcement on February 21, 1957, that the Dodgers had purchased the Los Angeles Angels from the Chicago Cubs had been headline news in the *Los Angeles Times*. The significance of the story was slightly less appreciated in New York, but the purchase did make the front page of the *New York Times*, where Roscoe McGowen described it as a "bombshell," a "surprising action [that] may have moved the Brooklyn Dodgers a little closer to becoming the Los Angeles Dodgers."[1] A major theme of McGowen's story was that New Yorkers hadn't appreciated how serious O'Malley had been in his pursuit of a replacement for Ebbets Field.

On balance, though, the acquisition of the Angels was seen simply as another "jab of the needle," as Arthur Daley had described the Jersey City gambit two years earlier. McGowen may have demonstrated unwittingly one of the reasons why New York officials were losing their hold on the Dodgers. Noting that the Dodgers needed the unanimous approval of the other National League teams in order to move to the West Coast, the *Times* reporter wrote, "the problem of such a vote will not arise for three years, since O'Malley has a lease on Ebbets Field for that period."[2] For all his perception in realizing that the Dodgers had established a tangible link with Los Angeles, McGowen failed to grasp the urgency with which New York must move in order to be sure it kept the team. The assumption that the city had three years to conclude an agreement with O'Malley lulled officials into adopting a leisurely approach in its negotiations.

"Leisurely" is the kindest description one can apply to the actions taken by New York officials to that point. The recurring hearings at the Board of Estimate, the various engineering reports, and the prolifera-

tion of administrative mechanisms to study the issue of a new stadium
are familiar characteristics of any major public works venture—bridges,
tunnels, and dams are notorious for their delays and cost overruns.
But the typical public works pattern was unacceptable to O'Malley, who
genuinely feared for the future of his club. The Brooklyn Sports Center
Authority was not especially inefficient compared to other administrative agencies; yet the money lost by repeated deferrals of action was
drawn not from the pockets of an amorphous group of taxpayers but
from O'Malley himself, who noted the delays and their attendant costs
and pursued other options accordingly. In its coverage of the Dodgers'
acquisition of the Angels, the *New York Times* discussed the effort of
the Brooklyn Sports Center Authority and concluded, "It has not looked
promising recently"; and Mayor Wagner reacted to the Dodgers' action
by stating, "We do not want to lose any of our baseball clubs. We will
do everything we can to keep them here." Brooklyn Borough President
Cashmore concurred, claiming he was "deeply interested in trying to
keep the major league Dodgers in Brooklyn."[3]

These statements reflected an attitude among New York officials that
proved decisive to the franchise's departure. Their bland rhetorical re-
plies to specific actions pushed the Dodgers inexorably away. At the
beginning of January 1957, weeks before O'Malley bought the Angels,
he had issued an ultimatum: "Unless something is done within six
months, I will have to make other arrangements. There is still a short
time left before we could be forced to take an irrevocable step to commit the Dodgers elsewhere."[4] O'Malley marked time in months; New
York officials thought they had years. In stark contrast to Wagner's and
Cashmore's vague reassurances, Los Angeles Mayor Poulson responded to the Angels' purchase by announcing he would fly to Vero
Beach within ten days "in an all-out effort to bring the Dodgers to Los
Angeles as soon as possible."[5]

In the wake of Poulson's visit to Vero Beach, New York writers split
regarding the likelihood of the Dodgers moving to Los Angeles. The
most astute observation read, "Los Angeles' case was strengthened if
only because six officials took the pains to come all the way to Florida
to demonstrate their sincerity to O'Malley. Since the Brooklyn-Chicago
deal was announced two weeks ago, all that O'Malley has gotten from
New York officials is a telegram and a phone call."[6] Not only Poulson
acted promptly in Los Angeles. City and county officials met on March
11 to determine what was needed to turn Chavez Ravine into the
Dodgers' future home.[7] The Ravine was the only site in Los Angeles
that interested O'Malley, and since the city had been eager for years to

develop the property, the conflict over a stadium location, which plagued New York for years, was not so serious an issue in California.

Similarly, administrative rivalries were averted in Los Angeles by the inclusion of the mayor and key members of the city council and county supervisors in the critical phases of decision making. At the March 11 meeting, the decision was made to proceed with a comprehensive study of financial, legal, and engineering issues that needed to be resolved before a stadium could be constructed for the Dodgers. The group also considered encouraging San Francisco to pursue a major league franchise since that city had already approved the construction of a municipal stadium.

While the effort in Los Angeles indicated a consensus among the key decision makers and their serious consideration of the appropriate issues, the activity in New York was about to become even more confused. On March 24, 1957, Abe Stark, the New York City Council President whose district represented Brooklyn, proposed a new stadium site for the Dodgers, the Parade Grounds, thirty-nine acres adjoining the southern end of Prospect Park. Stark estimated that a new stadium could be constructed for $7 million—$9 million if the Dodgers still wanted a weatherproof dome.[8] Stark chided O'Malley for conducting a "cold war" with the city over the stadium issue, then pointed out that his proposed location was served by major highways, two subway lines, and three bus routes. The site would give the Dodgers the modern stadium they wanted while keeping the team in Brooklyn, but the plan failed two critical tests: it met the approval of neither Moses nor O'Malley.

The indefatigable Moses, acting in his capacity as Parks Commissioner, rejected Stark's cost estimates for the stadium. The more accurate figure, according to Moses, was $10 million, even without the dome. He asserted further that 70 percent of the cost could be financed by the stadium's own revenues; the city would have to subsidize the rest.[9]

Moses's opinion was decisive, just as it had been for the Atlantic-Flatbush site. Moses was joined by O'Malley, who also rejected Stark's "disappointing" proposal.[10] Up to this point, O'Malley and Moses had joined forces in rejecting two possible stadium sites in Brooklyn—the Parade Grounds and one of the two options offered by the Clarke-Rapuano study of the Atlantic-Flatbush area. O'Malley continued to insist on his original goal, while Moses snubbed all ideas and offered no alternatives.

Mayor Wagner, meanwhile, declared he was confident the Dodgers would remain in New York. The mayor's Special Committee on the

Brooklyn Downtown Area Study, the administrative runaround devised by Moses the previous December, announced that a new engineering study would be made of the Atlantic-Flatbush site. This study would be conducted by the firm of Madigan-Hyland, which would replace the team of Clarke and Rapuano. Mayor Wagner assured the press that the Parade Grounds site would be among the options considered by Madigan and Hyland.

The Parade Grounds proposal was one of numerous sideshows in a more fundamental contest, the final confrontation between O'Malley and Moses. On April 18, 1957, Moses at last announced his solution to the Dodgers' problem. He proposed the construction by the Parks Department of a fifty-thousand seat stadium on the site of the 1939 World's Fair in Flushing Meadows, Queens.[11] The park would cost between $10 and $12 million, would include the often-mentioned dome, and would provide parking for twelve thousand cars. Moses's offer was important in two respects. First, it was the only realistic offer in New York for a replacement for Ebbets Field. All the other prospects—the Parade Grounds, the Atlantic-Flatbush sites, Staten Island, and Jersey City—were the doings of people without the power to effect a result. If Moses could persuade O'Malley to accept the Flushing Meadow location, the Dodgers would likely remain in New York. The second critical aspect of Moses's proposal was that by locating the stadium outside Brooklyn it removed sentiment from the issue. O'Malley had already noted that once the team left Brooklyn, it was no longer the Brooklyn Dodgers. The franchise was now free to pursue the best offer, rapidly narrowing to a choice between Flushing Meadows and Chavez Ravine. Moses's proposal seemed curious to some because he had previously followed a policy of keeping parkland intact. A disingenuous explanation was offered by a Parks Department official who declared that Moses himself was not interested in the Dodgers but that "the Wagner Administration was, and Mr. Moses works for the Administration."[12]

A more persuasive answer might be gleaned from Robert Caro's biography of Moses. Caro contends that Moses had long envisioned a kind of super urban park surpassing anything yet built, and the site for the park was Flushing Meadows:

> From the 1920's the Flushing Meadows had fired Moses' imagination. Part of the inspiration had been their size, of course; they were 1,346 acres, a Central Park and a half. Part had been their location: they lay almost precisely at the geographic center of New York; as the city's population shifted, moving steadily eastward, its population center was moving steadily closer to those meadows, too—a park there, Moses felt, would be a true "Central Park" to the whole city. For a man with a vision so broad that it

required vast open spaces for realization, here was a vast open space—at the city's geographic and population center. And in some way the very ugliness of the meadows seemed to furnish inspiration, too. They, and the Riker's dump beyond, were "a cloud of smoke by day and a pillar of fire by night," he was to write. Rereading Isaiah, he came across "Give unto them beauty for ashes"—after that, his dream had a slogan. He would turn what may well have been the ugliest part of New York City into its most beautiful.[13]

Moses first expected his dream to materialize after the New York World's Fair of 1939–40, and then after the second great World's Fair of 1964–65. Between those occasions, he tried to establish the United Nations headquarters at Flushing Meadows, and having missed that chance he offered the Dodgers inclusion in his dream. To most men the offer might have been enticing, but in Walter O'Malley, Moses was dealing with someone not interested in accepting a secondary role in someone else's dream.

O'Malley's reaction was cautious: "If and when this latest reported possibility ever achieves political maturity, I will be pleased to discuss it in detail."[14] Brooklyn Borough President Cashmore was less polite. He charged that a move to Queens would be a stab in Brooklyn's back, and added that Moses's suggestion was known to him only as a joke made by the borough president of Queens.

For his part, the Queens borough president, James Lundy, replied that he favored the plan because it would keep the Dodgers in New York without removing land from the tax roles, unlike the Atlantic-Flatbush proposals.[15] The burgeoning rivalry between Queens and Brooklyn is another indication of the fragmentation of authority in New York as more and more officials intervened with "solutions" to the stadium issue, and thereby introduced further confusion. Since both Lundy and Cashmore were members of the city's Board of Estimate, their competition would have to be resolved before any serious bid could be advanced to retain the Dodgers.

As April passed, the pattern continued: effective action from Los Angeles; confusion and rhetoric from New York. In the first week of May, O'Malley prepared to visit Los Angeles, ostensibly to look over the Angels and Wrigley Field. At the same time, Cashmore proposed a variation of the Atlantic-Flatbush plan in which the city would purchase the land while the Dodgers underwrote the stadium itself. The Board of Estimate approved another $25,000 to study the redevelopment of downtown Brooklyn, but only $5,000 were earmarked for the study of a new stadium.[16] Again at a critical point in O'Malley's decision making the Board of Estimate offered negligible support.

When O'Malley returned from Los Angeles, he found little changed in New York. In fact on May 16, Abe Stark brought the stadium issue full circle. Rebuffed in his attempt to use the Parade Grounds, he now proposed that the Dodgers stay in Ebbets Field, which would be expanded to accommodate fifty thousand fans—with parking for five thousand cars—through the acquisition of run-down land adjacent to the park.[17] O'Malley rejected the plan out of hand: "Mr. Stark continues to add confusion to what would have been a simple solution, had he given initial support."[18] A week after Stark's announcement, the mayor's Special Committee on the Brooklyn Downtown Area Study declared that the only two sites being considered for the Dodgers were the adjoining lots in the Atlantic-Flatbush area.[19] The committee ruled out Moses's Flushing Meadows proposal, the expansion of Ebbets Field, and the Parade Grounds. As might be expected, Moses's bid didn't die so easily; the Board of Estimate was considering it separately.

Since January 1957, when O'Malley had issued his six-month ultimatum, he had acquired the minor league property of the Chicago Cubs in Los Angeles, met with officials there to discuss their concerted plans for attracting the Dodgers to their city, and flown to California, where he was assiduously courted by city and county leaders who promised unwavering support. In New York, meanwhile, came irreconcilably diverse solutions from the president of the city council, the Brooklyn borough president, and Robert Moses. The only tangible backing was minimal funding for engineering studies by the Board of Estimate and Mayor Wagner's rhetorical endorsements of vague measures to keep the team in New York. In May, the Special Committee, which steered proposals through the administrative morass, announced in effect that the only serious proposals were those offered the previous October by the Clarke-Rapuano study, the very proposals which had attracted such meager support in December from the Board of Estimate, and which Robert Moses had opposed from the beginning.

As the fate of the Dodgers' franchise was being determined by public officials, the team that had dominated the National League in the 1950s was beginning its demise. In 1957, the Dodgers won eighty-four games, a respectable record, but one that dropped them to third place, eleven games behind the Braves and three behind the Cardinals.

Roy Campanella, the great catcher, was in the twilight of a career that would end with a nearly fatal car accident. Pee Wee Reese, Don Newcombe, and Jim Gilliam failed to produce as they had in previous seasons, while Duke Snider, Gil Hodges, and Don Drysdale finished among the league leaders.[20] The Dodgers were a good team, but others were better and had more gifted young players. None of the young Dodger hitters promised to replace Snider, Gil Hodges, or Furillo. Don Drysdale seemed the heir apparent to Newcombe, but the other pitchers gave little evidence of being able to win much—especially if the team's run production declined—although one young hurler, Sandy Koufax, who finished at 5–4 with a 2.88 earned run average, had a blazing fastball. A realistic appraisal of the future—and O'Malley was capable of no other kind—foresaw seasons of struggle.

While New York continued to dither about finding a new stadium site for the Dodgers, another hurdle was cleared in Los Angeles. National League owners voted unanimously on May 29, 1957, to authorize the Dodgers and Giants' move to the West Coast, as long as both clubs confirmed their intentions before October 1.[21] Walter O'Malley and Horace Stoneham both denied that the league vote indicated that their clubs would definitely leave, and both claimed the action merely opened the door to potential opportunities that each had been exploring independently for some time.

For the league owners authorization was a significant step because it indicated that prospects—not only for the two clubs in question but for others as well—were brighter in California than in New York, where the league had been founded eighty years before. (This would leave a vacuum which, rumor had it, could be filled by the Cincinnati Reds, who might leave their own small park, Crosley Field.)[22]

In response to the National League vote, Mayor Wagner called a meeting of all principals on June 4 for the purpose of taking decisive action to keep the Dodgers and Giants in New York. The session was advertised as a showdown, with all pertinent issues on the agenda, including television revenues, new stadium sites, and the possible use of Yankee Stadium. But like so many decisive meetings before, it resolved nothing. Platitudes were exchanged about good-faith efforts to keep the teams in New York. Wagner pledged to expedite the engineering reports and to discuss a final offer with O'Malley in late July.

This was a time of almost comical misperceptions by the two cities about their relative holds on the Dodgers. Norris Poulson felt desperate because O'Malley seemed in such a commanding position, while New York observers imagined an almost irresistible offer from Los Angeles, one that would actually have been impossible to effect. Before the June 4 meeting the *New York Times* reported that the mayor and the Board of Estimate had flatly declined to subsidize a baseball team, since in their view a franchise was simply another private business. The *Times* contrasted this attitude to what it thought was Los Angeles's offer "to spend $3,500,000 to give the Dodgers publicity and a stadium."[23] "Los Angeles desperately wants the Dodgers," Arthur Daley wrote, "and will go to almost any lengths to woo them to the Coast."[24] Peter Golenbock concludes that when O'Malley met with Los Angeles County Supervisor Kenneth Hahn at the 1956 World Series, "the fate of the Brooklyn Dodgers was sealed."[25]

This confusion is another of the factors that have contributed to the O'Malley Devil Theory. Its believers interpret all O'Malley's actions during the summer of 1957 as ruses meant to delude Brooklyn fans so that they would continue to attend games in Ebbets Field. The argument is presented most directly by Damon Rice: "Never was the owner of the Dodgers more devious or cold-blooded, than during the summer of 1957. He had already made up his mind to move to Los Angeles but continued to deceive the City of New York and the people of Brooklyn. He told us there was still a chance the Dodgers would stay long after the chance was gone. All he cared about now was squeezing the last nickel out of the Faithful."[26]

Evidence for this position is offered by Golenbock, who quotes Supervisor Hahn on O'Malley: "He said, 'I will come down here, but I will deny it to the press' because he had another season to play at Ebbets Field. He said the Dodgers' fans are rough fans. Literally would kill him, he said."[27] Even conceding the accuracy of Hahn's recollection, O'Malley's remark remains as far-fetched as Rice's interpretation.

The flaw in the conventional analysis is the assumption that O'Malley or Hahn or Poulson had the power to construct a stadium. As we have seen, the offer extended by Hahn was nullified by Arneburgh, and the subsequent agreement between Los Angeles and the Dodgers was constructed in spite of formidable legal, economic, and political barriers. At the time that Golenbock dates O'Malley's decision to shift the team, the engineers' report calling for a stadium in the rugged terrain at Chavez Ravine had not even been issued. Once the National League authorized the Dodgers' move, Rice despaired of their staying in Brooklyn; but at that point no mechanism had been devised to finance a stadium in Chavez Ravine, Poulson had reported to O'Malley that a municipal stadium was impossible, and Harold McClellan had not even been appointed to conduct negotiations with O'Malley.

Each city seemed to think the other held powerful sway over the Dodgers and that O'Malley occupied a lofty position, playing off his suitors. To be sure, the Dodgers' president operated craftily, but of all the participants he was perhaps the most vulnerable. He had sold his ballpark, had yet to receive a realistic offer for a replacement in New York, and could not be sure another would materialize in Los Angeles.

The Dodgers of the World War I era.

Ebbets Field shortly after it opened in 1913.

Dodger manager Wilbert Robinson confers with team president Steven McKeever.

Two men who led the Dodgers' resurgence in the 1940s: Larry MacPhail *(left)* and Leo Durocher.

June 15, 1938. The first night game in Ebbets Field, and Johnny Vander Meer's second consecutive no-hitter.

Ebbets Field and its environs.

Larry MacPhail with the championship team he assembled in the 1940s.

Walter O'Malley with some of the "Boys of Summer."

The antagonists: Walter O'Malley and Branch Rickey.

Brooklyn's world champions in one of the Jersey City games of 1956.

Robert Wagner, O'Malley, Robert Moses, and John Cashmore discuss a new stadium for the Dodgers in Brooklyn.

Jackie Robinson stealing home under Yogi Berra's tag in the 1955 World Series.

The final Opening Day in Ebbets Field.

ackie Robinson and Branch Rickey, their
attle won.

Walter O'Malley greeted upon his arrival in Los
Angeles by County Supervisor Kenneth Hahn
and Councilwoman Rosalind Wyman.

Mayor Norris Poulson officially welcomes the Dodgers to Los Angeles.

Gil Hodges scores the winning run over the Braves in the 1959 National League playoffs.

A record crowd of over 90,000 fills the Los Angeles Coliseum for the 1959 World Series.

Gil Hodges batting against Early Wynn in the fourth game of the 1959 Series.

Larry Sherry and Duke Snider celebrate a world championship in the transitional year of 1959.

Dodger Stadium under construction in Chavez Ravine.

AERIAL VIEW OF DODGER STADIUM
THE DAY IT OPENED—APRIL 10, 1962

(Herald Examiner Photo by Harold
Morby from KTLA-TV 5 Telecopter
piloted by Harry Scheer)

Dodger Stadium's inaugural, April 10, 1962.

Maury Wills stealing his record 104th base during the 1962 season.

andy Koufax after sweeping the Yankees in
963.

Walter Alston, who managed the Dodgers to
championships in Brooklyn and Los Angeles.

The O'Malleys, Walter and Peter, about 1962, with National League President Warren Giles.

The only place where the Dodgers could count on playing was a park in Jersey City that was significantly smaller than Ebbets Field.

A more plausible explanation for O'Malley's cautious manner is his awareness that in attempting to exchange a lucrative market for an even richer one, he could end up far worse than he had begun. To those who believe that by June 1957 the Dodgers had a lock on a new home in Los Angeles, O'Malley's actions that summer seem cynical and manipulative. The evidence reveals, however, that a new home for the Dodgers in California was far from inevitable and that O'Malley's maneuvers were those of a man caught between the tedious and unpredictable political machinations of two of the nation's largest cities. Self-interest was undoubtedly the motivation for O'Malley's behavior, but his intention was to make sure his franchise remained dominant rather than to wring the last nickel out of the Brooklyn fans.

Interestingly, Horace Stoneham never generated the hostility still felt toward the Dodgers' president. One reason Stoneham was spared may be that his family had owned the Giants since 1919, and had more credibility than O'Malley, an interloper who brazenly disrupted a love affair between a team and its borough. The Giant's move merits its own history, but some aspects of it are inextricably linked with the Dodgers' and need to be considered to fill out the Dodgers' case. On the face of it, the two teams confronted similar problems. The Polo Grounds, like Ebbets Field, was an obsolete park in an inhospitable part of New York, the northeast section of Manhattan across the Harlem River from Yankee Stadium in the Bronx. A housing project had been constructed adjacent to the ballpark, and, as in Brooklyn, little land was left for major renovations at the original site.

The 1950s Giants are generally depicted as virtual wards of the Dodgers. Their attendance relied on the interborough rivalry, so when the Dodgers left for the West Coast, the Giants necessarily had to follow. Twenty-four dates into the 1955 season, the Giants had drawn 513,476 fans, of whom 220,739 had attended seven games with the Dodgers.[28] During 1956, the Giants' home attendance had dropped to 629,267, and more than a third of those fans had attended the eleven home games between the Giants and Dodgers.[29] David Voigt offers the standard conclusion: "The ramshackle Polo Grounds and its deteriorating neighborhood no longer attracted crowds. Stoneham, facing certain bankruptcy, harkened to O'Malley's proposed west coast movement. Actually, Stoneham had no choice; New York writers knew this and poured most of their ire on O'Malley, while Stoneham prospered."[30]

The ballpark was certainly run-down, and attendance was in decline;

but neither condition necessarily dictated bankruptcy as the following data indicate:

	Attendance[31]	Net Income[32]
1952	984,940	−$222,344
1953	811,519	− $63,307
1954	1,155,067	$395,725
1955	824,112	$151,113
1956	629,267	$81,415

The ability of the Giants to sustain a profit in 1955 and 1956 in spite of dismal attendance and a fall in the standings to third and then sixth place was the result of television. In 1955 the Giants took in almost $650,000 from broadcast revenues, an amount second only to the Dodgers' and more than twice that of the third-place Cardinals.[33] In 1956, the Giants increased their receipts to $730,000, again far outpacing every other team except the Dodgers.[34]

The combined broadcast revenues for the Dodgers and Giants exceeded those receipts for the entire rest of the League. If the Dodgers left New York to the Giants, a lucrative broadcast market would probably have been the Giants' alone, and would have provided a secure financial base, along with a new stadium, possibly the one proposed by Robert Moses in Flushing Meadows. Since the Giants were not pinned with a borough identification, the shift from Manhattan to the more populous Queens would not likely have excited the wrath that greeted Moses's proposal to O'Malley. If this scenario seems familiar, it is because it was subsequently enacted by the Mets. Created as an expansion team in 1962, the Mets played for two years in the Polo Grounds, then moved to Shea Stadium in Flushing Meadows, next to the 1964 World's Fair site, where they have prospered, while the Giants, meanwhile, have battled to survive in San Francisco.

The Giants' lucrative year in 1954, when they won their last world championship, indicates that the franchise was far from hopeless when Stoneham decided to move to San Francisco. The team had struggled following its championship season; but with Willie Mays in center field, it was hardly doomed. Mays was the only true Giants star in their final years in New York, but Red Schoendienst played second base for them from mid-1956 to mid-1957, when he was traded to the Braves and helped them win the National League pennant. Had he been retained, the Giants' story might have been quite different, for reinforcements were on the way. In 1958, Orlando Cepeda began a distinguished career by winning Rookie of the Year honors for the Giants. The following year another Giant, Willie McCovey, was the League's best rookie, hitting 13 of his eventual 521 home runs. In 1960, the pitching staff received a needed boost with the addition of Juan Marichal, who would win 238 games for the Giants, more than any pitcher for the club in the modern era except for the immortals Christy Mathewson and Carl Hubbell.

These players formed the nucleus of the team that in 1962 took another pennant from the Dodgers in a playoff, and then lost a 1–0 seventh game to the Yankees in the World Series. That team remained competitive and challenged for pennants through the 1960s. They won a Western Division title in 1971, traded Willie Mays to the Mets in 1972, and almost every year since have struggled in the San Francisco fog. The Giants have changed owners several times, and contemplated relocating again in Denver, San Jose, or even Washington, D.C.

The Giants's fortunes have been so disappointing that one must reexamine the evidence available in the 1950s to determine whether it compellingly supported a move. In *New York City Baseball*, Harvey Frommer writes of the Giants' president: "Stoneham was not pushed. He jumped."[35] Stoneham himself told Peter Golenbock, "I had intended to move the Giants out of New York even before I knew Mr. O'Malley was intending to move. I was unhappy playing in the Polo Grounds. The ballpark was old, and it was darn near impossible to finance one in that area. I had intended to go to Minneapolis."[36] The attraction of Minneapolis was that the Giants owned that city's minor league franchise, and a new stadium had recently been constructed that could accommodate major league baseball. Also, because of its proximity to Milwaukee, Chicago, and St. Louis, Minneapolis would not represent a radical break in the dispersement of franchises, unlike the shift to the West Coast.

Chub Feeney, currently the president of the National League and in 1957 the Giants' general manager, confirms that the Giants were preparing to leave New York before they learned of O'Malley's designs on California. He contends that broadcasting revenues were only a minor consideration at that time, and that the Polo Grounds were increasingly problematic because of declining revenues.[37] The Giants, who owned the stadium, although the land it stood on was retained by the Coogan family, rented the facility to the New York Giants football team as well as to several college football teams. But by 1957 those teams had all abandoned the Polo Grounds in favor of other sites, such as Yankee Stadium, and the lost rent weakened the finances of Stoneham's operation. Feeney says that to his knowledge Robert Moses never offered the Giants the Flushing Meadows package that he extended to the Dodgers, and that the only specific proposal advanced to keep the Giants in New York was the bizarre suggestion by the Manhattan borough president, Hulan Jack, to construct a 110,000-seat stadium that would rest on stilts above the New York Central Rail Road yard on the West Side of Manhattan.

Another proposal for keeping the Giants in New York was offered by George McLaughlin, the former head of the Brooklyn Trust Company who had hired MacPhail, Rickey, and Walter O'Malley to restore the Dodgers to respectability. In 1957 McLaughlin worked for Robert Moses, as vice chairman of the Triborough Authority. On June 6 McLaughlin wrote to Horace Stoneham proposing an unprecedented plan.[38] On behalf of a non-profit organization, which would raise funds by public subscription, McLaughlin offered to purchase the Giants. He anticipated raising $5 million from "business corporations and public and civic minded citizens."[39] New York City would then build a stadium in Flushing Meadows, for which the Giants would pay rent instead of taxes. The $5 million could be used either to purchase the Giants from the Stoneham family or, if the Stonehams declined, to acquire another team that would represent the National League in New York.

The most radical part of the McLaughlin plan would have established a profit-sharing plan for the players. This particular feature was pure heresy because, as Branch Rickey put it, it "might get the players thinking they were part owners."[40] The baseball owners were still riding high as the lords of baseball, and profit sharing must have seemed another aspect of the Communist peril. McLaughlin's plan was rejected by Horace Stoneham, who replied in writing that the Giants were not for sale.[41] In a subsequent meeting with the Giants' owner and in some additional correspondence McLaughlin repeated the offer, with variations.

At the same time, McLaughlin wrote to Mayor Wagner to apprise him of the efforts to keep the Giants in New York. McLaughlin reported, erroneously, that Stoneham had invited the Cincinnati Reds to move to New York, thereby indicating that the Giants were committed to moving to San Francisco. Upon discovering his error, McLaughlin apologized to Stoneham and contended that, in fact, it was Walter O'Malley who had extended the invitation to the Reds.[42] The available evidence contains no proof of that allegation, but it may have reinforced the opinion among New York officials that as of June the Dodgers were committed to leaving but the Giants might yet stay. A memo dated June 13, 1957, and probably written by Moses considers the various possibilities of the two teams leaving or staying. No financial problems were anticipated in any eventuality, but if the Dodgers stayed and insisted on remaining in Brooklyn, the stadium issue would have to be resolved. The memo concludes, "We probably can solve that but I believe O'Malley is so far committed that he must go."[43]

On June 20, McLaughlin went public with his plan. He acknowledged that Stoneham had declined his offer to buy the Giants, and so the $5 million would serve to purchase another existing National League club or to acquire a new team through league expansion. Robert Moses heralded the ideas as the "first intelligent, honest, realistic appraisal."[44] Mayor Wagner sidestepped a position by stating that the plan was McLaughlin's, not his own, but that he remained committed to doing everything possible to keep both teams in New York.

Wagner's expressed hope was considered futile by the key actors. William Shea, an attorney who helped create the Mets, telegraphed Warren Giles, the president of the National League, to notify him of the "so-called Moses, McLaughlin plan to have a National League team in New York City in the event both the present National League teams leave the City."[45] McLaughlin supported Shea with a letter to Giles, dated June 24, in which he stated that he and his associates were "applying for a franchise in your league, the National, to be located in the Borough of Queens."[46] McLaughlin reviewed the advantages to the league of retaining a franchise in New York, and strongly emphasized that he and his associates were unwilling to wait until the end of the season before taking action to secure a replacement for the city. He expressed serious doubts that the Giants and Dodgers would have taken the steps they already had unless they intended to transfer to the West Coast. In McLaughlin's mind the "alluring offers" made by Los Angeles and San Francisco clinched the moves: "It looks as though these two cities are prepared to give terrific subsidies that no other city is prepared to meet."[47] McLaughlin didn't distinguish between the two West Coast offers, which differed significantly; and whether either should have been termed a "subsidy" is debatable. The central point, which McLaughlin keenly grasped, was that the political choices of New York and the California cities were the decisive elements in the fate of the teams:

> In New York City, if any give-away program such as has been mentioned by the Giants and Brooklyn was attempted, the taxpayers would rebel and even if the City authorities were inclined to give such an extravagant proposition, I am satisfied that it would be stopped through court action. However, there is nothing so far indicated that our City authorities are going to succumb to the unusual pressure that has been put on them with the threats of shifting franchises and removals.[48]

It is uncertain that the citizens of New York would have reacted so strongly if Robert Moses had made the Atlantic-Flatbush site available to the Dodgers. It is equally unclear that the courts would have sustained such a use of urban renewal authority. What is beyond dispute is that key elected and administrative officials in New York were disinclined to use their power to accommodate the Dodgers and Giants. One can speculate endlessly about the wisdom of these officials as well as their motives; the fact remains that they could have taken measures to keep the two teams in New York but chose not to.

The contract with Los Angeles couldn't have been more striking. On June 11, the city Housing Authority, the agency that had sold Chavez Ravine to the city in 1952, announced that in response to Mayor Poulson's request they would change critical language in the deed of sale that might have presented an impediment to the use of the land for the construction of a baseball stadium.[49] The *Los Angeles Times* reported the announcement: "Commission Counsel James J. Arditto explained that the mechanics of ironing out the deed probably would involve nothing more than a rewording or amplification of the deed's language in conferences between authority representatives and the Public Housing Administration which has remained a party to the deed."[50]

In the end, the Chavez Ravine agreement would receive a full airing in the highest courts as well as in a ballot referendum. But before it reached that stage, the perceptions and actions of officials in New York and Los Angeles were critical. In New York, Robert Moses interpreted the Federal Housing Act as precluding the use of its slum-clearing power for the construction of a baseball stadium. In Los Angeles, officials recognized potential problems with the same law but concluded they could be solved with minor rewording of an existing agreement. Again, the key distinction was not the relative power of the public officials, but their willingness to use what power they had to pursue their interests. The nature of those interests was also important.

In June 1957, the controversy attending the Dodgers and Giants reached the United States Congress, where the antitrust subcommittee chose to reconsider one of the subjects dearest to the hearts of the baseball owners: the game's exemptions from antitrust law. What had been long regarded as an anomaly in the law was first established in the 1922 Supreme Court case, *Federal Baseball Club of Baltimore, Inc. v. National League of Professional Baseball Clubs, et al.*[51] The case concerned a suit brought by a defunct Federal League club seeking damages on the grounds that the agreement between the National and American leagues amounted to a restraint of trade in violation of the Sherman Antitrust Act.

The Court's opinion, written by Oliver Wendell Holmes, held that baseball was exempt from antitrust law because it failed to qualify as interstate commerce, which alone is subject to federal regulation. Justice Holmes acknowledged that teams crossed state lines for commercial purposes, but he wrote that "the transport is a mere incident, not the essential thing." "The business is giving exhibitions of baseball, which are purely state affairs."[52] This opinion, fanciful as it seems, was more realistic than a New York State Supreme Court opinion in 1914 which concluded that "Baseball is an amusement, a sport, a game that comes clearly within the civil and criminal law of the state, and not a commodity or an article of merchandise subject to the regulation of Congress."[53]

At the end of World War II, the financial complications that arose from the challenge of the Mexican League, television, minor league claims, and player unions patently put the lie to the judicial rulings of a more innocent time. In 1951, Congressman Emanuel Celler, a Democrat from Brooklyn, convened hearings before his antimonopoly subcommittee of the House Judiciary Committee in order to reconsider baseball's exemption from antitrust law. These hearings resulted in a useful history of practices such as the reserve clause and the minor league draft, but the ensuing report, printed in 1951, led to no legislation proposing that baseball conform to one of the most basic legal constraints on American business. James Dworkin has written that Congress took no action because it wished neither to confirm baseball's antitrust exemption nor to remove the sport completely from the internal control of the reserve clause.[54] Furthermore, baseball owners assured committee members that pending litigation would determine the legality of the reserve clause without the need for legislative action.

The owners were correct in a sense, but they neglected to mention that they would fight with all their resources to keep the reserve clause. Their opportunity to do this arose in the next term of the Supreme Court in the case of *George Toolson v. New York Yankees, Inc.*[55] Toolson, a member of the Yankees, rebelled against his demotion to the club's farm team in Binghamton, New York and refused to report there. He was subsequently placed on a list of ineligible players and barred from playing anywhere in organized baseball. Toolson sued for triple damages on the grounds that the reserve clause constrained trade, in violation of antitrust laws. A seven-member majority of the Court dispatched the case in one paragraph that demonstrated the Alfonse-Gaston manner in which courts and legislatures often interact. The court noted its earlier opinion in the *Federal Baseball* case, and then wrote, "Congress has had the ruling under consideration but has not seen fit to bring such business under these laws by legislation having prospective effect."[56] At a later point it stated, "We think that if there are evils in this field which now warrant application to it of the antitrust laws it should be by legislation."[57] Toolson lost because in 1922 the Court had assumed baseball was exempt from antitrust law, and in the ensuing thirty years Congress had taken no action to suggest that this opinion was wrong. The Celler committee had deferred to the courts, which in turn deferred to Congress, and a judgment resting on an outdated interpretation of interstate commerce remained in force.

In February 1957, baseball's privileged position was emphasized by the Supreme Court's ruling in *Radovich v. National Football League* that professional football did not enjoy the same exemption from antitrust law that baseball did.[58] The explanation was simply that, although baseball had developed for thirty years under the *Federal Baseball* decision, football had never received such a specific exemption. The limited wisdom of the Court's continued exemption of baseball from antitrust law isn't worth deciphering. The Court relied on the concept of *stare decisis*, which can be interpreted here to mean that the mistake made in 1922 had been accepted by all affected parties without being remedied by those with the authority to do so. Tom Clark's majority opinion in *Radovich* included another invitation to Congress to correct the longstanding anomaly: "We, therefore, conclude that the orderly way to eliminate error or discrimination, if any there be, is by legislation and not by court decision."[59] After a paean to the wonders of the legislative process, Justice Clark wrote, "Of course, the doctrine of *Toolson* and *Federal Base Ball* must yield to any congressional action and continues only at its suffrance."[60] As recently as 1972 the Supreme Court reaffirmed this reign of error in the Curt Flood case. The majority specifically upheld baseball's exemption from antitrust law and football's inclusion within those laws: "If there is any inconsistency or illogic in all this, it is an inconsistency and illogic of long standing that is to be remedied by the Congress and not by this Court."[61]

Such a remedy was precisely what Congressman Celler wished to consider when he reconvened his antimonopoly (now renamed "antitrust") subcommittee of the House Judiciary Committee in June 1957. The specific invitation of the Supreme Court in the *Radovich* case was to reexamine baseball's use of the reserve clause, a power found illegal when employed by football teams. But the 1957 hearing spent little time on the reserve clause. The committee focused instead on franchise relocation, specifically the proposed moves of the Dodgers and Giants. In Celler's first question to Warren Giles, the president of the National League, he asked, "Let's go right into it. Let's put the fat right in the fire. What do you know about the contemplated or intended removal of the Giants and Dodgers to the west coast?"[62]

Giles replied that he had heard a lot of "idle chatter" about it and that to minimize speculation he had encouraged club owners to meet in May to air their reactions to a prospective move by the New York teams. Giles contended that he himself took a neutral position, although he reminded the owners that the Dodgers had been drawing over a million fans a year to Ebbets Field. When asked whether O'Malley had made his request for permission to move on the basis of financial considerations, Giles responded that he had not, that he had argued instead for the need for a new ballpark because Ebbets Field had been sold. (No one mentioned that O'Malley was essentially asking for relief from a predicament he had initiated.)

Celler reached the heart of the matter when he reviewed the Dodgers's profits for the preceding five years. He asked Giles, "Do you think it meet and proper for the club to ask the city fathers to build a stadium for them?"[63] Giles answered that O'Malley had intended to build the stadium himself after the city helped the club acquire land at an appropriate site. Celler then focused on the manner in which O'Malley wanted to obtain the land: "He wanted to exercise what we call in the law the right of eminent domain to condemn the property, which is not usually done except for what we call public convenience and necessity."[64] At that point, Kenneth Keating, a Republican later elected U.S. Senator from New York, interjected, "Would not that be convenient and necessary for the Dodgers to continue in Brooklyn?" Celler replied, "Maybe. It all depends upon your point of view."[65] This point was then dropped but, of course, it remains the essence of the entire controversy. In New York, the point of view was that a baseball stadium did not merit the use of governmental authority to acquire needed land; in Los Angeles city and county officials believed a major league stadium fit the legal requirements for the use of Chavez Ravine.

From his perspective as league president, Giles insisted he could not say definitely whether the Dodgers would leave for Los Angeles or stay in Brooklyn: "I can put myself in their position. They don't know whether all of the obstacles that have to be overcome will be such that they will go."[66] This remark was a suitable prelude to Walter O'Malley's testimony before the committee. Celler opened with what he described as the burning question, "Can you tell the committee at this time whether or not the Dodgers will play in Los Angeles next year?" "I'm sorry," O'Malley responded, "I cannot answer that question."

CELLER: Why?
O'MALLEY: I do not know the answer.
CELLER: Why do you not know?
O'MALLEY: I do not know the answer for two reasons. One, I do not know
          what the result of Mayor Wagner's study in New York City
          will bring. Two, I do not know whether or not Los Angeles
          will be ready for major league baseball next spring.
CELLER: What preparations have you made for moving to Los Angeles?
O'MALLEY: None whatsoever.[67]

O'Malley's last disingenuous remark was then explored at length. Celler introduced a memorandum that O'Malley had brought voluntarily to the hearing. Written by the Los Angeles city attorney, Robert Arneburgh and dated May 3, 1957, it listed ten points that would form the basis of an agreement between the city and the Dodgers for accommodating the franchise on the West Coast. Celler asked O'Malley if he had accepted the terms of the memorandum, and O'Malley replied that he had not. Celler pressed for a reason, and O'Malley dissembled in a manner worthy of Casey Stengel: "Well at the time that was prepared it was an idea. The idea would have to come into fulfillment by certain of those things being done. It was prepared not only on the potential that the Dodgers might move to Los Angeles, but that, on that basis, some major-league club would be interested in Los Angeles."[68]

Celler then asked if O'Malley wasn't playing Los Angeles and New York against each other, and O'Malley denied the charge. Celler asked for elaboration and O'Malley began a review of the history of his efforts to find a new stadium for the Dodgers. He dated those efforts from 1947 and said the Dodgers had hired an engineer and had discussed the matter with New York City officials. O'Malley stressed his objective that the new stadium be part of a larger urban redevelopment plan so that it would be eligible for government help in securing the land through the power of eminent domain. The Atlantic-Flatbush site should qualify under such a design, O'Malley contended, because the Fort Greene meat market would be moved, lowering the price of meat in Brooklyn five cents per pound, and because traffic jams at the intersection would be eased through new traffic patterns. Other benefits would derive from added parking and the removal of substandard dwellings in the area. Finally, the Long Island Rail Road would receive its new depot and track to accommodate its new rolling stock. "And all of this" the Dodger president concluded, "would have magically left enough acres of land on which a ball park could be built, at the cost of the owners of the Brooklyn Ball Club, not one penny of which was to be paid by the city of New York."[69]

After noting that Robert Moses ruled out the use of eminent domain under Title I of the Federal Housing Act, O'Malley turned to the option of the Brooklyn Sports Center Authority. He reviewed the extensive political support he had received in lifting that proposal over various legislative hurdles. At one point, he said, Moses had indicated that he wanted to know how much the Dodgers themselves would contribute to buying the Authority's bonds if the agency were established. O'Malley replied that the team would spend $4 million. Moses had challenged O'Malley to demonstrate he could raise such a sum, and O'Malley responded that, if the agency were established, he would produce the $4 million. After the bill had passed the legislature, Governor Harriman had told O'Malley he would come to Brooklyn to sign the bill into law and assured O'Malley he was a Dodger fan and supportive of the objectives of the new law. O'Malley told Celler, "So I was doing pretty well, Mr. Chairman. I was winning all the battles. The only thing I lost was the war."[70] O'Malley then covered Mayor Wagner's appointments of the three members to the Sports Center Authority, and recounted his efforts to raise the $4 million he had earlier promised he could produce. "Well, I have a one-track mind and go right down the line, right down the middle, so the first thing I did was to sell Montreal for $1 million. I then sold the Brooklyn Baseball Park for $3 million. . . ."[71]

Celler asked why O'Malley had secured only a three-year lease from Marvin Kratter, the new owner of Ebbets Field. O'Malley replied, "Oh, this show was on the road. This was something where we were going to start throwing up steel overnight. It looked to me like my whole work of 10 years was finally going to blossom out into the finest baseball stadium any place in this country."[72] At this point in his testimony, O'Malley introduced the erroneous point that he consulted with the mayor of Jersey City to provide a backup in case the new stadium were not constructed in the three-year period of the lease back. In fact the announcement of the arrangement in Jersey City had been made in August 1955, before Ebbets Field was sold. Indeed, seven games were played in Jersey City during the 1956 season before the sale of the stadium and before the Sports Center Authority was established. O'Malley later corrected his statement, but did not explain why he erred, nor did he address its effect on his testimony. Because so much has been made of his alleged desire to mislead the people of Brooklyn, one might note that the error was so blatant and so easily discovered that it makes a sinister motive for the misstatement unlikely. In any case, whatever the implications of this part of O'Malley's testimony, Ebbets Field was sold and Los Angeles had taken no direct action to promote a baseball stadium in Chavez Ravine and give O'Malley an alternative to staying in New York.

O'Malley next turned to the December 1956 decision by the Board of Estimate to provide the Sports Center Authority with a fraction of the budget the agency had requested. "I was very much chagrined to find that when the Sports Center Authority went in for a budget so they could have their studies made in keeping with the timetable which Mr. Moses had carefully prepared, that Mr. Stark at that time showed dissent in the matter and he said, 'What Brooklyn needs more than a new stadium are two legitimate theaters and an opera house.' "[73] The Dodgers' president then indulged in a personal attack on Abe Stark, stating at one point, "Now he is a swell little fellow but he doesn't know what this is all about."[74] The basis for O'Malley's animus toward Stark appears to have been the council president's support for a 5-percent admission tax imposed by New York City on sports and other entertainments. O'Malley first reminded Celler that the congressman had joined the Dodgers in opposing the tax when it was first imposed, then added, "Last year the Brooklyn Dodgers paid $495,000, that was the cream off the top of the bottle, in admission taxes, and by gosh, I think that is disgraceful because that it is a tax 'other businesses' do not pay."[75]

O'Malley's testimony on the Board of Estimate's action missed two key points. First, he overestimated Stark's influence. As city council president, Stark sat on the Board and, as a citywide official, he held two votes. Stark, however, had proposed two options for the Dodgers in 1957—the Parade Grounds site and remodeling Ebbets Field. Their practicability notwithstanding, the proposals showed that Stark was not opposed to a replacement in Brooklyn for Ebbets Field. The second point O'Malley missed had to do with Moses's attitude toward the Board of Estimate vote. While not a member of the Board, Moses nevertheless was extremely influential and, as we have seen, he encouraged Mayor Wagner to back a much lower budget for the Sports Center Authority than that sought by its chairman, Charles Mylod.

Later O'Malley and Celler engaged in badinage about the Dodgers' profits. O'Malley crowed that he was "very proud indeed" of his earnings, and that "it is a good old American custom and I am glad I am doing well." Celler replied, with obvious sarcasm, "I glory in your profits. . . . I hope you make twice as much next year."[76] O'Malley then spoke at length about the Sports Center Authority and its funding by the Board of Estimate, charging that the Authority had been "sabotaged" by people who voted for the agency only because they were sure it would never win passage in Albany. When the Authority did receive legislative approval, "they weren't willing to back up their vote with money for appropriation."[77]

O'Malley provided no evidence for these allegations, but the charge is certainly plausible. Public officials commonly lend overt support to measures popular with the electorate. If powerful interests then oppose the measure, those same officials delay or nullify the project. It is unclear whether the Sports Authority fell prey to such shenanigans, but its efforts were undoubtedly hampered by the meager support it had received from the Board of Estimate.

Representative Keating asked O'Malley directly whether—had New York provided adequate funding for the Authority the preceding year—the Dodgers would still contemplate a move to the West Coast. O'Malley replied that he had never spoken to anyone in Los Angeles "or any other darn place about the Dodgers ever leaving Brooklyn until I was given very excellent reason to believe that this thing was not going to be processed in Brooklyn, and euchred myself unfortunately in a position where I had sold my ball park and I had to have a place in which to play ball. . . ."[78] O'Malley went on to praise the Los Angeles officials: "The first group of people that came up and showed that they had a little old-fashioned American initiative and what not were the people from Los Angeles, and they showed that they had political unanimity out there, the Republicans and the Democrats and the publishers of the papers, and they said, 'We are for you, and when you tell us you don't want us to build you a ball park, that is most refreshing and amazing.' "[79]

The last statement diverges somewhat from the actual record since, according to Norris Poulson, O'Malley was at least testing the waters about a possible public stadium for the Dodgers in Los Angeles. At any rate, O'Malley concluded with a ringing call for private enterprise: "I have never asked the city of New York to build me a ball park, to give me land, to give me a subsistence or a subsidy, nor have I asked Los Angeles or anyplace to do it. I don't want to be a tenant in a political ball park, I want to own my own ball park and run it the way I think it should be run."[80]

Chairman Celler objected to some of the insinuations about New York public officials, and O'Malley replied with a perceptive comment about their problems: "Mayor Wagner has been most cooperative, Mr. Chairman. He however is the mayor of the whole city and he can't get into a political hassle just over Brooklyn. . . . If the man in Queen's [sic] won't vote for this to be done in Brooklyn and the man in the Bronx won't vote for it, then Mr. Moses gets disgusted with the whole thing and says let's put them over in Flushing Meadows, that is what we have got."[81]

O'Malley recounted a conversation he had with Moses after the Dodgers had purchased the Los Angeles Angels and the final offers from New York were being proposed. According to O'Malley, Moses invited the Dodgers' president to his home, and asked, "Walter, is this a hopeless cause? Can anything be done?"[82] O'Malley replied that he was still unsure whether Los Angeles would be able to put a deal together. If New York could finally overcome its problems, it was still possible to keep the Dodgers there. Moses inquired about Abe Stark's proposal for the Parade Grounds. O'Malley said he rejected the idea because of his ostensible concern for the lads who played sandlot baseball on the site. More compellingly, he added that parking was inadequate and rapid transit connections poor. O'Malley had then offered to buy Fort Greene Park near the Atlantic-Flatbush site, but Moses blocked that proposal because the park was seriously needed by residents. The commissioner then extended the Flushing Meadows offer and invited O'Malley to look at the proposed site. O'Malley agreed, but with serious misgivings that again reflected an awareness of the political problems he faced in New York. By relocating the stadium in Queens, O'Malley recalled, "I would get the votes from Queens but I would lose probably the votes from Brooklyn, and I still wouldn't necessarily pick up the votes from the Bronx and Manhattan, nor would I have the controller's vote."[83] Nevertheless, when Moses asked O'Malley if he would return the city's capital investment in a new stadium in Flushing, O'Malley claimed he would. This point is in dispute, however. Abe Stark had testified that no response was received from O'Malley, but O'Malley contended that after sending his letter of acceptance he never heard from Moses again on this matter.

Keating and Celler tried to pin O'Malley down about whether he would still stay in New York if the proper replacement for Ebbets Field were provided. O'Malley evaded several queries because he deemed them too hypothetical, although he did indicate that his purchase of the Angels did not commit the Dodgers to a move. The banter continued as O'Malley eluded questions about the Dodgers setting up shop in Queens or in a stadium whose specifications were left vague. Finally, Celler found the precise wording: "Let's assume that the city fathers give you the facilities that you want, a 50,000 seat stadium, parking space for a minimum of 4,000 cars, despite the fact you have a $2 million investment in Los Angeles, would you remain in Brooklyn?" The answer: "Yes."[84] Such straightforwardness couldn't last. As Keating began another inquiry, O'Malley interjected, "But time is running out on that."[85] Further questions indicated that O'Malley intended to make a decision at the end of the 1957 season.

Keating, a Republican, asked O'Malley questions that seemed intended to present the Democratic administration in New York in the least favorable light. He asked O'Malley to confirm Keating's impression that the Dodgers' president had agreed on two occasions to the city's plans for the construction of a new stadium, and that on both occasions the city had changed the terms. The first instance was the Atlantic-Flatbush proposal of the Sports Center Authority, and the second was the Flushing Meadows site proposed by Robert Moses. O'Malley indicated that Keating's impression was correct. O'Malley praised Wagner and Cashmore, but took another swipe at Abe Stark's idea for remodeling Ebbets Field. He than jousted with the committee counsel about whether the Dodgers were bound for Los Angeles, discussed at some length the Skiatron venture in pay television, and concluded with some general observations on the application of antitrust law to baseball.

Celler's committee had pressed O'Malley and Warren Giles to determine if the Dodgers were moving to Los Angeles. Both men replied that they did not know. In O'Malley's case especially, this answer was greeted with some skepticism, yet the reply rings true. Before the Dodgers could commit to any city for the 1958 season, they had to have the stadium issue resolved. In Los Angeles, O'Malley had found public officials willing to try to solve his problems, something missing from New York, but their good intentions had not yet translated into a firm offer to provide land for a new ballpark. Conceivably, if Los Angeles had not been able to make the Chavez Ravine offer, O'Malley might have settled for something like Moses's Flushing Meadows project. At the time of the hearings, O'Malley was pessimistic about staying in New York because no possibility seemed to lead to the stadium he had wanted at the Atlantic-Flatbush site.

O'Malley needed a firm and tangible offer from at least one of the two cities. At the time of the hearings, he had nothing to bank on from either. The committee counsel asked, "What is it that the mayor or the city council or Mr. Stark can show you between now and the end of this season that will give you justification for staying in New York?" O'Malley replied, "The Atlantic and Flatbush Avenue solution, which is the only thing I have been fighting for."[86] O'Malley added, "Things are moving very rapidly and very intelligently in Los Angeles. . . . I told them that I thought the jig was up in Brooklyn and they are pick-

ing up and doing some pretty smart work out there right now."[87] If
O'Malley was deceiving anyone by saying he was not yet sure where
the Dodgers would play in 1958, it was himself.

In mid-July, Horace Stoneham, president of the Giants, testified before Celler's committee that he preferred the Giants to move to San
Francisco if that city offered a suitable plan.[88] Its mayor, George Christopher had, in fact, discussed such a move with Stoneham several times,
and the city had passed a bond resolution to construct a stadium for
any major league team which transferred to the Bay Area. The day
after his testimony, on July 18, Stoneham made two further announce-
ments.[89] First, he declared there was still a slim possibility that the
Giants could stay in New York if the city constructed an appropriate
facility and charged a reasonable rent or if the Giants were permitted
to use Yankee Stadium. But again Stoneham doubted the Giants would
be welcome across the Harlem River. Second, he announced that wher-
ever the Giants were located in 1958, they would not be playing in the
Polo Grounds. Stoneham pointed to declining attendance, poor parking, and increasing traffic jams as the basis for abandoning the stadium
that had housed the Giants since 1911. Stoneham also extended a sympathetic word to Walter O'Malley, saying that the Dodger president "is
getting a kicking around from the press and the politicians in a[n elec-
tion] year."[90]

This sympathy for O'Malley was not shared by Robert Moses, who
wrote an article for the July 22 issue of *Sports Illustrated* rehearsing his
own version of the stadium issue.[91] He charged that "For years, Walter
and his chums have kept us dizzy and confused," and further alleged
that O'Malley's "first substitute Brooklyn site proposal was nothing if
not nervy and Napoleonic."[92] The article discussed the Atlantic-Flatbush
proposals in a similar vein, Moses again attacking the site favored by
the Clarke-Rapuano study without mentioning that O'Malley also preferred a different parcel at the Atlantic-Flatbush location. Moses also
raised the applicability of Title I, claiming that "From the point of view
of constitutionality, Walter honestly believes that he in himself constitutes a public purpose."[93] Moses muddled the issue by chiding O'Malley
for failing to consider larger recreational uses for an all-purpose stadium. In fact O'Malley had considered such a proposal, but he had made
no secret of his preference for privately owning a stadium exclusively
designed for baseball. Reviewing the history of the Sports Center Authority, Moses predicted the expected report from the Madigan-Hyland
engineering firm would be to no avail since delays and indecision on
the part of the city about the various components of the plan had left
the Authority "a dead duck." The Stark proposal for the Parade Grounds

was dismissed by Moses for lack of public support, bringing him to the "only suggestion involving park land which makes any sense."[94] That "suggestion," of course, was the Flushing Meadows site. He reviewed the advantages of the McLaughlin proposal, noted Brooklyn fanaticism and also the "slim attendance" at Ebbets Field, and concluded with a call for his all-purpose stadium.

O'Malley dismissed Moses's article as "an infanticidal attack on the Brooklyn Sports Center Authority which he helped to procreate."[95] He also distanced himself from Horace Stoneham's statement concerning a move to the West Coast, announcing that he stood by his testimony of the month before that he would keep the Dodgers in Brooklyn if a suitable alternative to Ebbets Field could be found.[96] The Atlantic-Flatbush site received another blow in August with the release of the engineering report by Madigan-Hyland.[97] The firm studied the location preferred by Walter O'Malley as well as the area favored by the Clarke-Rapuano report of 1956. A stadium at either site was estimated to cost over $20 million, and when the necessary improvements to the area were included the total cost escalated to about $50 million.

"It is evident", the report commented, "that such an expenditure cannot seriously be considered for the sake of a stadium alone. Its only justification is a large-scale rehabilitation scheme for the general area."[98] And there would likely be further complications: exacerbated traffic congestion, unprofitable parking facilities, and a wait of at least four years while the project was completed. The *New York Times*, reporting the firm's findings, concluded that the prospect for a stadium at the Atlantic-Flatbush site "neared the vanishing point yesterday."[99] On August 19, in a letter to John Theobald, a deputy mayor who had worked on the stadium issue, Moses noted that housing would have to be a part of any general rehabilitation of the Atlantic-Flatbush site, although "It is very doubtful whether there will be enough federal writedown money available to add a project at this place."[100]

Curiously, the Madigan-Hyland report appeared not to discourage Walter O'Malley, who interpreted its findings as a death blow to the Sports Center Authority and thus as a vote for his original plan to purchase the land and construct the stadium privately. On August 26, O'Malley reiterated his first offer for keeping the Dodgers in Brooklyn. He met with Deputy Mayor Theobald and proposed that the city condemn land at Atlantic and Flatbush and sell it to the Dodgers, who would then construct their own stadium and pay taxes to the city on both the land and the facility. The *New York Times* incorrectly reported that this offer "for the first time stressed that the Dodgers were willing to buy the land, build the stadium and become full-fledged taxpayers."[101] The error suggests how pervasive the assumption was that Walter O'Malley wanted the city to build a stadium for the Dodgers.

The proposal coincided with trouble in Los Angeles—some members of the city council were balking at the Chavez Ravine proposal. Harold McClellan eventually mollified those members, but their resistance may have encouraged O'Malley to try New York once again. A memo from a meeting of the Committee on the Sports Authority mentioned that O'Malley proposed to purchase the land himself and build a stadium "estimated to cost $7,500,000 and [that he] would pay full taxes on both the stadium and the land."[102] An additional study of this proposal was scheduled to be presented to the Board of Estimate on September 16. The memo stated that the Dodgers would have to declare their intentions to the National League by midnight of September 30, and that "at least three clubs are very much interested in the Los Angeles location."[103] The memo suggests that in spite of all obstacles Walter O'Malley was still willing to try to stay in Brooklyn according to his original design; it also indicates that Los Angeles was not in a desperate position since other teams were interested in the area if the Brooklyn franchise were unable to shift.

John Cashmore remarked that O'Malley's proposal "raised certain legal questions which would have to be explored by the Corporation Counsel."[104] Perhaps coincidentally, the counsel, Peter Brown, received a letter the next day from Robert Moses, who expressed his views on O'Malley's design: "I hope you won't give the time of day to Walter O'Malley's latest 'offer.' The Atlantic Avenue site is dead for a Sports Center. Time, delays and other factors have killed it. . . . Acquiring land for sale to Walter O'Malley is not a public purpose and would be a scandalous procedure anyway. The Authority is out of business and can't finance anything, whatever its original possibilities may have been." "What remains," Moses pronounced, "is Flushing Meadows."[105]

Moses was undoubtedly correct that by August 1957 the options had been whittled away. Funding for Title I projects had diminished significantly, and delays had indeed devastated the Sports Center Authority. More significant, however, was the expression of unwillingness on Moses's part to assist this final effort to keep the Dodgers in Brooklyn. The same implacable opposition that had been present from the start remained to the very end, blocking the efforts of those who sought to accommodate O'Malley's designs to stay in Brooklyn.

The memo from the mayor's committee and Moses's letter further belie the commonly accepted assumption that O'Malley was set to move to Los Angeles many months before the final announcement. Based on available information, things were very much in flux as late as August 1957.

While the Dodgers' status remained in doubt, the Giants announced on August 19 that they were moving to San Francisco after the 1957 season. "We're sorry to disappoint the kids of New York," Horace Stoneham said, "but we didn't see many of their parents out there at the Polo Grounds in recent years."[106] A twelve-point agreement between the Giants and San Francisco provided for the construction of a municipal stadium. The proposal was made to the Giants in a letter of intent from the city's mayor, George Christopher.

Warren Giles declared that the league resolution of May, which seemed to permit a franchise shift to the West Coast only if both the Giants and the Dodgers left, should not be so construed: that Giants' move did not dictate a shift by the Dodgers. In spite of Stoneham's announcement, the move may not have been as definite as first appeared. On August 29, Jay Coogan, whose family owned the land on which the Polo Grounds stood, wrote to Robert Moses that he had spoken with the Giants' lawyer, who had told him that the move was no more settled than the Dodgers'. "I am informed by the Chairman of the Board of a San Francisco bank," Coogan said, "that a move to California by the Giants under the plan so far promulgated and made manifest by the Mayor of San Francisco and the National Exhibition Company cannot possibly eventuate."[107]

Arthur Daley's column in the *Times* of August 23 considered the Dodgers' position in the wake of the Giants' announcement. He noted that by acting noncommittally O'Malley gave "the hot foot to lagging Los Angeles and . . . spurs civic officials in New York to strain a bit extra to keep him. Only the elastic O'Malley can ride two horses in opposite directions at the same time."[108] Daley also noted that "With the Giants irrevocably gone, the Dodgers now have the National League segment of the rich New York territory all to themselves—if they want it." He speculated that the Flushing Meadows site might now be more attractive to O'Malley because of its central location and eventual pay television prospects. After offering several interpretations of O'Malley's actions, Daley remarked, "This entire business gets curiouser and curiouser."[109]

The final month of the 1957 season saw the final efforts by both New York and Los Angeles to secure a commitment from the Dodgers. While O'Malley was on a hunting trip in Wyoming, Louis Wolfson, a financier, offered to buy the club for $5 million and keep it in Brooklyn.[110] On September 9, the *New York Times* carried a story which indicated that even as Los Angeles offered increased acreage at Chavez Ravine, opposition there was mounting. Anonymous "skeptics" were quoted disparaging Los Angeles's ability to sustain a baseball team and the city was described as "a place where the sidewalks are rolled up by 7 P.M."[111]

Wolfson's offer to buy the Dodgers was never a serious possibility, but on September 10 another offer was extended from someone who was always taken seriously. Nelson Rockefeller announced that he had tried for a month to work out a plan for keeping the Dodgers in Brooklyn.[112] He indicated he was willing to become part-owner of the franchise or to aid in the construction of a stadium which could then be used for other purposes, such as youth recreation. The announcement discouraged Mayor Poulson, who commented, "If it is true that Mr. Rockefeller has entered the picture, I'm very much afraid we don't have much of a chance to get the Dodgers. We want to make 'angels' out of the 'bums'; but we can't be Santa Claus like some of these big names."[113] Rosalind Wyman also evinced some pessimism, saying that Los Angeles might lose out "because of the way negotiations were conducted."[114]

New York's possibilities brightened again because of a legal opinion from Corporation Counsel Peter Brown. Rejecting Robert Moses's advice, Brown declared that the land at Atlantic and Flatbush could be acquired by the city for resale if the city Planning Commission determined that the area was substandard and unsanitary.[115] This would avoid the Title I authority and also circumvent Moses's powers. Deputy Mayor Theobald announced that the Board of Estimate would consider Brown's suggestion at its next meeting, on September 19.

That morning, O'Malley met with Rockefeller and Mayor Wagner for the purpose of seeing if Rockefeller could offset some of the financial losses the city anticipated from the condemnation of land for the stadium. The meeting preceded an executive session of the Board of Estimate, and after talking with Wagner and Rockefeller O'Malley agreed to delay any decision about a move until the Board rendered a decision on the Rockefeller offer. The details of the proposal, leaked to the press by members of the Board, called for the city to condemn the twelve-acre site for $8 million and then sell it to a corporation formed by Rockefeller for $2 million.[116] The corporation would then lease the site to the Dodgers, rent-free, for a period of twenty years. The Dodgers would construct a stadium on the location and pay all real estate taxes. The Board of Estimate rejected the plan, characterizing it as a giveaway of taxpayers' money. The specific objection was to the disparity between the $8 million cost of condemnation and relocation of established businesses and the $2 million Rockefeller would pay to the city. The Board apparently did not consider the other economic consequences of the

move, such as the potential impact of baseball attendance on businesses in the vicinity of the new stadium.

Robert Moses, meanwhile, continued lobbying against any help for
the Dodgers. In a letter to William Peer, Mayor Wagner's Executive
Secretary, he stated, "As the matter stands, Nelson [Rockefeller] has
been badly advised and the City has nothing to gain from an attempted
acquisition and sale of the Atlantic Avenue triangle to Walter O'Malley
which, among other things, would wreck the 1958 and 1959 Capital
Budgets if the facts were honestly disclosed. Nothing remains but the
Flushing Meadow proposal and we in the Park Department have lost
interest because of lack of support and the better chances of other projects. This Dodger business reminds me of the jitterbug jive."[117]

Moses's letter included a copy of a memorandum to Peter Brown in
which Moses offered some language to soften the suggestion from the
corporation counsel that the city could condemn land for the building
of a stadium. Moses said that he had informed Brown by phone that
"I think you will get an awful shellacking and that your reputation will
suffer in the end unless you add something like this to your opinion
or weave it into the opinion before you reach the end. To be quite
honest about it, I don't think this case is on all fours with your quote."[118]

On September 20, the Rockefeller plan collapsed when modifications
designed to placate the Board of Estimate—including Rockefeller's offer
to up his purchase price to $3 million—proved unacceptable to O'Mal-
ley.[119] O'Malley contended that under the terms of the proposal, if the
Dodgers exercised their right of first option to buy the land, the cost
would soar from $3 million, including interest, to $4.5 million, an increase that priced the Dodgers out of the plan because the franchise
would be paying real estate tax for twenty years.[120]

The refusal of the Board of Estimate to support the Rockefeller plan
ended New York's last opportunity to hold on to the Dodgers before
the Los Angeles City Council acted on the Chavez Ravine proposal.
The Board's action received the enthusiastic endorsement of Robert
Moses. "Of course I heartily applaud the action by the Board of Estimate in refusing a giveaway program to retain the Dodgers," he wrote
to Mayor Wagner, repeating a familiar request. "Why not have another
look at the idea of a Municipal Stadium at Flushing Meadow?"[121]

As we have seen, Los Angeles did not miss the opportunity afforded
it by New York. Taking advantage of the National League's extension
of its October 1 deadline, O'Malley waited until the city council approved the Chavez Ravine agreement on October 7 and the next day,
during a World Series game between the Yankees and Braves, the
Dodgers announced they would move to Los Angeles for the 1958 sea-

son. The manner of the announcement showed little consideration for the Brooklyn fans being left behind. A publicist for the Dodgers read the following statement to those in the press room at the Waldorf-Astoria Hotel:

> In view of the action of the Los Angeles City Council yesterday and in accordance with the resolution of the National League made Oct. 1, the stockholders and directors of the Brooklyn Baseball Club have today met and unanimously agreed that necessary steps be taken to draft the Los Angeles territory.[122]

Walter O'Malley was not in attendance for the announcement, nor was Brooklyn invited to bid its team farewell. A business decision made for sound if not compelling reasons was executed without regard for the emotional response it was sure to generate.

7

Proposition B

Brooklyn Dodger fans have occasionally been described as having been so worn out by the wrangling over a new stadium that they were indifferent when the news of the Dodgers' move was announced. A more perceptive description was offered by Arthur Daley, who characterized the mood as one of "galling resentment."[1] Daley wrote, "Other teams were forced to move by apathy, or incompetence. The only word that fits the Dodgers is greed." This expression of frustration suggests the powerful emotional grip that a baseball team can have on a community. Daley's remarks also reveal blindness to an essential aspect of the national pastime. "Baseball is a sport, eh? . . . the crass commercialism of O'Malley and Horace Stoneham of the Giants presents the disillusioning fact that it's big business, just another way to make a buck."[2]

The feeling of betrayal still persists among fans who rooted for the Dodgers in Ebbets Field.[3] The sentiment is understandable, but it shouldn't obscure the fact that since the inception of professional leagues business calculations have been among the sport's paramount concerns. They dictated the trading or selling of such stars as Babe Ruth, Ty Cobb, Christy Mathewson, Cy Young, Grover Cleveland Alexander, and more recently Willie Mays, Frank Robinson, and Pete Rose. Financial interests were crucial to such sportsmen as Connie Mack and Branch Rickey.

O'Malley's crime was to remove the fig leaf from baseball. He openly considered displayed what others had camouflaged in rhetoric. Thus, while his peers—Rickey, MacPhail, and George Weiss—have been elected to the Hall of Fame, even the suggestion that O'Malley belongs there would inspire fierce controversy. The picture of O'Malley as a scheming Machiavellian is totally at odds with the Dodgers' naive behavior after leaving Brooklyn. Not only were they set back by the mounting challenge to the Chavez Ravine deal, they had to scramble even to secure a temporary playing field pending the construction of the new stadium.

By December 1957, the Dodgers' popularity in Los Angeles had been demonstrated by the advance sale of tickets in excess of $750,000, the highest mark ever achieved in Brooklyn.[4] The club's vice-president, E. J. Bavasi, echoed his boss's reaction, noting happily that the new fans were paying in full rather than on installment, which had been the practice in New York.

But the Dodgers soon faced a serious threat to their new prosperity. On December 1, 1957, Walter Peterson, the Los Angeles city clerk, announced that enough valid signatures had been collected on petitions to place the Chavez Ravine agreement on a public referendum to challenge the validity of the contract.[5] The stadium issue would be on the ballot for June 3, 1958, enabling the electorate of Los Angeles County to pass judgment on the wisdom of the pact between O'Malley and city officials. The referendum was entitled Proposition B. A "yes" vote would confirm the agreement, while a "no" would nullify it. The referendum, as we have seen, was instituted by the Progressive movement as an electoral check on the abuse of power by public officials. This civics lesson would catch Walter O'Malley by surprise. "I never anticipated a referendum," he later remarked. "In fact, I was completely unaware of the thing they called a referendum because they never had that in New York. Very few places have it. They have initiatives and referendums out here. Very peculiar. No boss . . ."[6]

O'Malley's father had been a public official in New York City, and the Dodgers' president had spent his life around powerful figures in business and government. He had grown accustomed to wielding power behind the scenes, making quiet deals, and abiding by their terms. Now he discovered that the most important deal he had ever made could be reversed through an entirely open and public exercise of power. Private meetings with mayors, appearances before congressional committees, and bantering with reporters were political exercises which O'Malley handled deftly; he was now forced to appeal directly to the people.

He could blame only himself for his predicament. A suitable review of the pertinent facts of California government would have attuned O'Malley to the possibility of an electoral reaction to his move, just as the Chavez Ravine site was available in part because of the public referendum that challenged the construction of public housing on the site in the early 1950s. Even if he was confident about the outcome of the referendum, O'Malley must have chafed at the thought that work on the new stadium would be delayed; at least six months would pass before a final resolution of the stadium issue would be settled, which complicated the task of locating an interim site for the Dodgers' games

Unquestionably, the new developments left O'Malley unsettled. As the major leagues' winter meetings began in December, rumors developed that the fondest hopes of Flatbush might be realized; the Dodgers might return to Brooklyn.[7] The other National League owners were prey to such gossip because they had cooperated in abandoning the New York market, and were now discovering that the anticipated wealth of California might remain locked in politics and litigation. The basis for the rumors was the fact that the only certain place for the Dodgers to play in Los Angeles in 1958 was Wrigley Field, a minor league stadium worse than Ebbets Field. Its capacity was only twenty-three thousand, and there was virtually no parking. Two possible alternatives were the Coliseum, which required $500,000 in alterations, or the Rose Bowl in Pasadena, another facility designed for everything but baseball.

In the face of the confusion, O'Malley confounded things further by offering one of his open-ended denials. Insisting that the Dodgers were in Los Angeles to stay, he attempted to clarify an earlier statement that had stirred the waters:

> I think where my words were misconstrued was when I was asked whether it is still possible for the Dodgers to return to Brooklyn. The answer to that is yes. Because there still is a lot of paper work to be done before it's all signed and sealed officially. So that if anything should happen to make it impossible for us to open in Los Angeles we could still return to Brooklyn. . . .
>
> But as of now I have no intention of returning to Brooklyn. I have every hope the Dodgers will be able to do all of the things we want to do in Los Angeles, including the building of a stadium. I have the utmost confidence that the people of Los Angeles will show their displeasure at those special interests which are behind the referendum.[8]

Perhaps the reference to a return to Brooklyn was another of O'Malley's jabs, meant to spur local officials to action.

As late as January 1958, the Dodgers were expected to play in the Rose Bowl.[9] This shift was thought necessary because of the inadequacies of Wrigley Field and also because of snags involving the Coliseum. The Dodgers were being offered a design for that stadium in which the baseball diamond would be located at the eastern, or peristyle, end of the structure. This arrangement required batters to stare directly into the sun during afternoon games, and also necessitated the construction of temporary stands to close off that end of the stadium, which opened after a short embankment of seats. O'Malley wanted the diamond located at the closed western end, which would make life miserable for the right fielder but presumably improve run production. Objections to O'Malley's plan were raised by the three football teams who used the Coliseum—the Los Angeles Rams and the teams of the University of Southern California and UCLA all complained that a baseball infield would damage their turf.[10] Objections centered also on the Dodgers' proposed rent. The football teams paid 10 percent of their gross receipts to the city, and received no money from concessions and parking. The Dodgers proposed to pay 5 percent of their gross receipts, keep the concessions income—limited by an ordinance prohibiting the sale of beer or other alcohol at the Coliseum—and yield the parking revenues. They would pay for cleanup after their games. O'Malley typically characterized his terms as extremely generous, since even at 5 percent the Dodgers might pay the city far more rent than any other team in baseball if attendance reached the record levels commonly anticipated.

A stalemate in negotiations for the Coliseum inspired O'Malley and Mayor Poulson to consider other facilities. The Rose Bowl was an obvious alternative, since it was used relatively infrequently, and O'Malley began discussions with Pasadena officials to determine the costs of renovating the stadium for baseball. But many residents objected to the prospect of a heavy volume of automobile traffic disrupting the tranquility of the area, and on January 7, 1958, Lee Paul, a local attorney, announced a petition drive to put another referendum on the June 3 ballot blocking baseball at the Rose Bowl if the city signed a contract with the Dodgers.[11] Pasadena's city attorney claimed such a referendum would probably prove illegal because it could have no retroactive force on a contract expected to be signed any day.

All these snafus rattled some lords of baseball. Commissioner Ford Frick, appearing with Red Smith on a television program in New York, opined that National League owners were beginning to feel they had made a mistake in allowing the move to the West Coast.[12] He went on to describe Wrigley Field as a "cow pasture" whose minor league dimensions undermined the integrity of the games played there. For this rare foray into blunt speech, Frick was rebuked the next day by Walter O'Malley, who declared that the issue of a stadium for the Dodgers was being resolved, and that baseball in Los Angeles would be legitimate even in a temporary facility.[13] The defense of Wrigley Field appeared crucial because on January 13, the deal for the Rose Bowl fell through. O'Malley promptly asserted that another effort would be made to secure a site in the Coliseum, although he anticipated the Dodgers would be playing in Wrigley Field until a new facility was completed.[14]

O'Malley's travails were the source of delight for many New Yorkers, including Larry MacPhail, who denounced the move to Los Angeles as the greatest mistake in the history of baseball, and who also offered to bet that the stadium in Chavez Ravine would never be built.[15] Support for O'Malley came from Philip Wrigley, who stated that the Cubs had developed plans for the expansion of Wrigley Field to another ten thousand seats at a cost of less than $50,000.[16] O'Malley added that the dimensions of the park could be expanded to be more suitable for the major league game.

During the week of January 12, the Coliseum Commission held hearings on a proposal for baseball in the west end of the stadium. Familiar objections were heard, along with the financial concerns of brokerage firms selling bonds for the construction of the Sports Arena, an indoor facility adjacent to the Coliseum. The brokers argued that any adjustment in the contracts between the city and the football teams could jeopardize the sale of bonds for the Sports Arena.[17] On Friday, January 17, the commission voted 9–0 to approve baseball in the Coliseum for the Dodgers.[18] The agreement covered two years, sufficient time, it was thought, for a new stadium to be constructed. The city anticipated it would receive over $600,000 in rent from the Dodgers during those two years, and temporary adjustments were made in the football contracts during that period. Kenneth Hahn, an early advocate of using the Coliseum as a temporary home for the Dodgers, points out today that the stadium was empty three hundred days a year, so that even contractual terms favorable to the Dodgers, who had to pay rent only on their first ten games, represented additional income for the city, county, and state.[19] These first few months in the era of the Los Angeles Dodgers demonstrated that far from being the smoothly calculated conspiracy depicted by its enemies, "The move," in the words of Harold Parrott, "was about as well thought out as a panty raid by a bunch of college freshmen who'd had too many beers."[20]

At a cost of $200,000, the Coliseum was converted into a monstrosity with the redeeming feature that it could accommodate crowds in excess of ninety thousand. Those crowds would see a field which ran 251 feet from home plate to the left-field foul pole. At that juncture, the notorious 40-foot screen rose as a bastion in defense of Babe Ruth's home-run record.[21] The screen extended 140 feet to left center field, where the power alley was 320 feet from home. Perhaps to provide a glimpse of the game to fans in the far eastern end of the stadium, the center-field fence was placed 425 feet from home and the right-field power alley an unbelievable 440 feet away. If Ruth's record fell to a left-handed hitter, it would have to be because of inside-the-park home runs. Controversy arose immediately around the league among pitchers, who anticipated numerous cheap home runs to left field. The Dodgers invited comparisons with Fenway Park and the Polo Grounds, but no one took them seriously. As a place to play baseball, the Coliseum was a joke. It was acceptable for two reasons: the fans and players could look forward to an ideal stadium in a couple of years; and meanwhile the owners could laugh all the way to the bank.

Harold Parrott describes some of the impediments to the fans' enjoyment:

> The sides of the Coliseum slanted away from the playing field like some gigantic saucepan. As a matter of fact, we *were* frying our customers alive in this stone skillet, but that wasn't all: The holders of the best box seats had to descend twenty-four rows below street level, which means they had to climb back up forty-eight steps to use the rest rooms or escape when the game was over.
>
> There were no elevators in the place then, and a climb to the cheaper pews really taxed the heart; by the time a fan reached the sixty-fifth row, his pump would have excited any cardiologist.[22]

The seats were wooden benches with straight backs. The capacity of over one hundred thousand-plus may have indicated that the average football fan is of narrower girth, for Parrott reports that on closer inspection the Dodgers reluctantly decided to remove a number of seats so that fans could be accommodated more comfortably on the hard benches. Padding was also added to the seats in a further concession to comfort.

O'Malley may have agreed to coddle fans because the only way they would be able to see the Dodgers was in person at the Coliseum. On January 19, O'Malley announced that, while all the Dodgers' games would be broadcast on radio, none would be shown on television.[23] Boosting the gate was one likely consideration; another was a more enduring interest—pay television. One week after the Dodgers announced their shift to Los Angeles in October 1957, the Los Angeles City Council approved a pay television franchise to Skiatron, the firm O'Malley had signed with in Brooklyn.[24] O'Malley remained the staunchest advocate of pay TV among major league owners, and he seized on the opportunity to benefit from such a system by broadcasting Dodger games in Southern California.

As with the Chavez Ravine contact, these designs were complicated by political and legal haggling. The new technology posed a significant economic threat to conventional television stations as well as to movie studios and theaters, which all sought to retard the new industry. One means of accomplishing this was to determine which organ of government held jurisdiction over pay TV. The Federal Communications Commission, established in 1934 to regulate the interference of competing broadcasters in the same spectrum space, seemed to have no authority because pay television uses a cable. Conflicts also arose between Congress and local governments eager to grant cable franchises. Congress preferred to move slowly and determine a national policy rather than leave the matter up to local interests.[25]

In Los Angeles, Skiatron executives found themselves further obstructed by the same device that was complicating matters for the Dodgers. The contract extended by the city council in October was challenged through a petition that placed the issue of pay television on the June 3 ballot. As Norris Poulson observed in the Dodgers' case, "Los Angeles is probably America's softest touch for a guy clutching a petition. A writer once observed that one putting his mind to it could gather enough signatures in the city to outlaw orange juice."[26] At the time of O'Malley's announcement about broadcasting, both the new stadium and the projections of pay TV receipts were in abeyance, yet to be determined by the voters. Perhaps O'Malley risked antagonizing people by denying them televised games, but he also risked losing revenues from commercial television contracts, receipts that had been vital in the prosperity he had enjoyed in Brooklyn during the 1950s.

The decision represented a sharp departure from the strategy O'Malley had pursued in New York. In a city he shared with two other teams, any decision that either the Dodgers, Giants, or Yankees made concerning television necessarily considered what the other two teams were doing. In Los Angeles, where he enjoyed a monopoly, O'Malley could pursue his interests without regard for the competition, and he clearly chose to support the efforts of pay television.

The harmony of interests between the Dodgers and Skiatron had been clarified during 1957 when Matty Fox, the president of the pay TV firm, discussed his future plans for the company. Fox appeared before the Celler committee in its hearings on organized baseball and testified that Los Angeles could be wired for pay TV by April 1958 at a cost of $12 million.[27] Such a service could not have been made available for Dodgers baseball in New York because the team had existing television contracts that would have remained in force through the 1959 season.

Fox also discussed the owners' incentives for using pay TV in an interview with the *Los Angeles Times* in July 1957: "The way it is now, the revenue baseball is getting from TV costs more in lost attendance than what the sponsors are giving them. Baseball people have the alternative of either dropping sponsors and cutting off television (for example, Milwaukee, with no TV had 700,000 greater attendance than the televised Yanks) or having their revenue supplemented by pay-TV."[28] Fox anticipated that the Dodgers could receive "several million dollars a year" from a cable system in Los Angeles. These revenues would not only be significantly higher than the broadcast returns enjoyed in New York but, presumably because of the cost of viewing, they would be less likely to reduce the gate attendance.

The attraction for O'Malley of the pay television system is obvious, but since both Skiatron and the Chavez Ravine contract faced uncertain electoral outcomes, each venture had to be handled with great care. An important distinction was that the team was very popular, while the television system was not. Opponents of the referendum on the Chavez Ravine contract addressed only the terms of the pact itself, not the desirability of having the Dodgers in Los Angeles. Skiatron, on the other hand, faced the insuperable problem of appearing to require payment for what previously had been enjoyed free of charge. An additional complication for Skiatron was the well-financed opposition from existing television stations and movie theaters. Part of Skiatron's programming would simply have introduced new sources of entertainment, such as opera, to a limited group of subscribers; but the firm also intended to broadcast highly popular features—Dodgers games as well as football games.

Faced with a losing outcome, Skiatron called for the city to rescind its contract unless the forthcoming referendum was postponed from the June to the November ballot.[29] When that postponement was not granted, the city council agreed to abrogate the agreement by a 10–1 vote.[30] Two weeks later, on April 27, O'Malley announced a new television policy for the Dodgers.[31] Henceforth, road games against the Giants would be televised back to Los Angeles. This meant that all games with the Giants could be seen in Los Angeles. The plan included no additional televising nor any prospects for another pay TV system in the near future.

The expressed rationale for this policy change was that the Dodgers had received numerous requests for televised games from shut-ins, especially those in Veterans hospitals.[32] No doubt such requests had been received, but the timing of the decision encourages a more hard-headed explanation. O'Malley may have taken into account two additional considerations. One was that the outcome of Proposition B on the Chavez Ravine referendum was now in serious doubt. Norris Poulson noted that polls had swung from 70 percent favorable to a slight majority opposed.[33] By making the Dodgers more accessible to the electorate,

O'Malley may have been hoping to swing voters to his side. A second possible consideration was that the team had stumbled from the gate at the start of the 1958 season. At the time of O'Malley's notice about televising games from San Francisco, the Dodgers' were 5–7, in sixth place in an eight-team league, three games behind the first-place Giants. Their position was hardly irretrievable, but their performance was doing little to generate excitement.

Since their move to Los Angeles, the Dodgers have displayed a relatively conservative policy toward televising games. With the exception of critical games during pennant drives—guaranteed sellouts—no home games have been broadcast over regular stations. Only road games from San Francisco were commonly televised until the 1970s, when Sunday road games were added. Later, the policy became somewhat more generous, expanding to fifty telecasts, but these too were always confined to road games until current cable systems put some home contests on the air. In contrast to many other teams and to their own policy in Brooklyn, the Dodgers have used television as a means of keeping their fans interested enough to encourage attendance in person when the team is home. The games in Los Angeles were preserved for the paying customer in a manner even Branch Rickey would have endorsed.

Perhaps the most formidable asset which Walter O'Malley might have brought to bear in support of Proposition B was the Dodgers' announcer, Vin Scully. Scully was immediately popular in Los Angeles, and with the team doing so poorly he became the focus for much of the enthusiasm about the team. He was the only member of the Dodgers having a good year. Scully relates that Walter O'Malley never discussed the political contest with him, nor suggested that Scully try to sway voters during his broadcasts.[34] Such an attempt might have been received as a crude measure and triggered a backlash, but the tactic was apparently never even considered. Scully also reports that he personally felt no responsibility for the outcome of the referendum. He maintains he didn't consider the seriousness of Proposition B until after the election was over. Up to that point, he was as unaware as Walter O'Malley about this exercise in direct democracy, and only later realized how fragile a hold the Dodgers had on Los Angeles. The announcer's innocence served the Dodgers very well.

As the major league teams prepared in Florida and Arizona for the forthcoming season, City Councilman Holland continued to lead opposition to the stadium contract. In March he called for a congressional probe into the agreement "so that all of the facts will be made available to the public."[35] Holland provoked spirited rejoinders from the Dodgers' supporters on the city council, especially Rosalind Wyman, who attacked him for issuing "misleading" statements about the contract.[36] Roger Arneburgh, the city attorney, also rebuked Holland for alleging that secret deals had been made between the city and the Dodgers during the negotiations of 1957.

The discussions became increasingly contentious, but no action resulted, in part because the Dodgers retained the overwhelming support of the city council and also because the June 3 referendum would definitively resolve the controversy. Holland's motion for a congressional inquiry was referred to a council committee, where it could be expected to languish, but the council did agree unanimously that if a congressional investigation ensued, it ought to identify the interests behind each side of the referendum. This last point echoed the argument of the council majority that the referendum had been sponsored by the owners of the San Diego Padres of the Pacific Coast League, who viewed the Dodgers as an economic threat. And indeed, although the campaign against Proposition B was couched in terms of resisting powerful interests in favor of a few beleaguered families living in the Ravine, the spokesman for the opposition turned out to be John Smith, the Padres' owner.[37]

While the preelection posturing continued, the Coliseum underwent renovation and the Dodgers prepared to wrest the National League title back from the Milwaukee Braves, a formidable task since Roy Campanella was definitely lost to the club as he gallantly fought for his life after his car accident. In addition, Pee Wee Reese had hit only .224 the season before and at forty could not be expected to contribute greatly in the new season. Duke Snider was recovering from knee surgery, and faced a right-field fence that seemed to reach the Mojave Desert. And Gil Hodges, who had a fine year in 1957, would probably be converted into a singles hitter by the left-field screen.[38] Younger players such as Jim Gilliam, Charlie Neal, and Gino Cimoli had played well the year before in Ebbets Field, but it was uncertain whether they could adjust to the radically different Coliseum. Untried players at critical positions—particularly catcher John Roseboro and third baseman Dick Gray—must be counted on heavily.

In 1958 the Dodgers introduced some players who would eventually achieve fame in Los Angeles but who were a few years away from major league status. Don Demeter, Ron Fairly, Frank Howard, and Norm Larker were outfielders who made their debuts that first season. Infielders Jim Gentile and Bob Lillis dreamed of replacing Brooklyn legends. Perhaps the face most familiar to Los Angeles fans was that of Steve Bilko, who had appeared in the major leagues in 1949 for the St. Louis Cardinals. Bilko had hit twenty-one home runs for the Cardinals in 1953 but never more than four in any other year. He remained a hero in Los Angeles, however, for his record power-hitting for the Angels in the Pacific Coast League.[39] Bilko was a classic example of the minor league superstar who could never quite make it in the big leagues.

The pitching was also problematic. Don Newcombe had fallen from his 27-win Cy Young Award season in 1956 to 11–12 in 1957. He had won 123 games in his first seven seasons, and at thirty-two was the most prominent member of the staff. In spite of these figures, Newcombe never achieved the success predicted for him and at the start of the 1958 season he was a major question mark. Don Drysdale had won 17 games in 1957, and at twenty-two seemed the future ace of the staff. Born and raised in Los Angeles, he also was a likely candidate to be the first star of the team's new era. Carl Erskine had compiled a 5–3 record during the last year in Brooklyn; arm problems severely limited his starts. Johnny Podres had won 12 games during the preceding season and led the league in earned run average. Roger Craig, Ed Roebuck, and Danny McDevitt were young pitchers being carefully groomed, as was Sandy Koufax, a "bonus baby" whose contract wouldn't allow him to be sent to the minors for some sorely needed fine-tuning. Young pitchers are always a gamble, but even established Dodger hurlers couldn't be considered sure bets for the 1958 season. To be sure, they had proved themselves in a hitters' paradise in Brooklyn; but the tactics they would have to employ in the Coliseum called for talents no pitcher had to demonstrate outside of Fenway Park in Boston.

The National League schedule had been revised to account for the two West Coast teams, and the Dodgers and Giants were slated to open their seasons against each other in two three-game series. The first meeting was April 15 in San Francisco's Seals Stadium, the former home of the local Pacific Coast League franchise. The Giants pounded the Dodgers 8–0, Drysdale taking the loss.[40] The next day the Dodgers rebounded as Duke Snider and Dick Gray hit home runs to pace a 13–1 victory, and Johnny Podres picked up the Dodgers' first win on the West Coast. The Giants won the series the next day, 7–4, despite a home run by Gino Cimoli. Don Newcombe was the loser.

On April 18, a Friday, the Dodgers held their home opener, following a tremendous display of civic hoopla. A new league opening-day attendance record was set as 78,672 fans endured a hot smoggy afternoon to see the Dodgers take a 6–5 win from the Giants. Dick Gray's home run broke in the Coliseum for the locals, and Carl Erskine was credited with the victory. On Saturday, the Giants drubbed Danny McDevitt 11–4 in spite of home runs by Gil Hodges, Charlie Neal, and Gray. This was the first night game played on the Coast and attendance was 41,303. In Sunday's finale, the Giants again belted Dodger pitching, 12–2, as Don Drysdale took his second loss without a victory. The Dodgers were left with a 2–4 record and impressive revenues— 47,234 attended the final game with the Giants.

The Dodgers continued to flounder, losing two of three to the Chicago Cubs. Their mediocrity inspired a boffo performance by the city council, which passed a 10–0 vote of confidence on April 25. Councilman Ernest Debs lamented that "We, who have brought them here, have stabbed them. These ball players are young men with no place for their families to live. How can they make commitments when they don't know whether next season will find them here or in Houston or Minneapolis—or back in Brooklyn?"[41] Pat McGee, who had joined John Holland in opposing the Chavez deal, objected that his position on the issue did not indicate a lack of support for the team. The final resolution expressed confidence that the team would be able to overcome its poor start and add luster to the splendid record that it had achieved in its final years in Brooklyn.

All this support did not improve the Dodgers' play. On May 12, the Giants drubbed the Dodgers for the seventh time in a month, dropping Los Angeles into the National League cellar with a record of 9–16, seven and a half games behind Milwaukee. The next day the Giants again pounded the Dodgers, 16–9. For the second time in as many days, Willie Mays hit two homers as the Giants continued to solve the puzzles of the Coliseum more successfully than their hosts.

On June 1 the Dodgers' fortunes improved momentarily when rookie Stan Williams two-hit the Cubs to win 1–0. This was the last game before the referendum. In what could hardly have been worse timing, the Dodgers were last in the National League with a record of 17–26, ten games behind the first-place Giants.

The bright spot continued to be the support of the fans. The weekend after the referendum, the Dodgers achieved one of the few triumphs of the season when they swept a three-game series from the Braves and drew over 50,000 each game. After twenty-seven home games, the Dodgers had attracted more than 800,000 fans, and they were more than 150,000 ahead of the Milwaukee Braves' record pace of 1957, when they drew 2,215,404.

Midway through June, with the team going nowhere, Don Newcombe was traded to Cincinnati for Johnny Klippstein and Steve Bilko. The Dodgers hoped that Bilko would recapture the magic of his Coast League days in Los Angeles, and help an anemic hitting attack which was last in the league with a .253 average. But Bilko was unable to reverse the team's performance. The Dodgers completed the first half of the season in last place and with the lowest batting average in the league. Snider and Furillo continued to hover around .300, but injuries limited their power to a combined 13 home runs and 65 r.b.i.'s. At the All-Star break, Charlie Neal led the club with 14 home runs and Gil Hodges had added 12; but so few runners were getting on base that Neal and Hodges together had only driven in 62 runs. The lowest earned run average among the starters was Williams's 3.46, while Drysdale, who had been expected to be the new ace, carried an earned run average of 5.66. Sandy Koufax had an impressive 7–3 record, but his e.r.a. was 3.48, and he was on his way to a club record 17 wild pitches for the season.[42]

More encouragingly, Don Zimmer and John Roseboro, the replacements for Pee Wee Reese and Roy Campanella, were both batting .272 with 6 home runs. After a good start, Dick Gray had fallen off, but he he still had 9 home runs. The team had turned a league-leading 113 double plays, almost 30 ahead of the Cardinals, who were next best. A realistic assessment of that accomplishment, however, would have to note the contribution of a pitching staff that was allowing far too many base runners.

As the season paused for the All-Star game, the Dodgers were 33–42, in last place, eight games out of first. The only consolation was that, except for the Giants and Braves, no team was performing with much more success. A brief spurt of better than .500 ball lifted the Dodgers to a sixth-place tie with Pittsburgh on July 21, and the two teams were only four games out of third. The Dodgers, however, were much closer to the cellar, to which they returned after the Pirates defeated them three straight times.

August supplied the Dodgers' best stretch of the year. After a loss to the Braves, Los Angeles continued their mastery of the Reds, winning three in a row and extending their streak against Cincinnati to nine before they dropped the nightcap of a doubleheader. A momentary rise to seventh place was reversed by two defeats to the Cardinals. The mediocrity continued until August 13 when the Dodgers defeated the Cubs 6–5 before some 14,000 home fans. The following evening the smallest crowd of the year, 6,195, saw the Dodgers climb out of the cellar again with a 7–3 win over Chicago. The poor performance of the other teams was apparent on August 15 when by sweeping a doubleheader from the Cardinals the Dodgers vaulted into a tie for fourth place, their headiest status of a dismal campaign. Two days later they split another doubleheader with St. Louis, and remained tied with them for fourth.

On August 19, the Dodgers took a doubleheader from the league-leading Braves, 4–1 and 7–2, before more than forty-two thousand in the Coliseum. On the following night, Stan Williams outpitched Warren Spahn for a 2–1 victory, the Dodgers' seventh in their last eight games. They had taken eleven of seventeen games from Milwaukee and were the only team with an edge over the defending champions. Lew Burdette hurled a 4–0 shutout over the Dodgers in the fourth game of the series, but Don Drysdale led the Dodgers back in the finale, with a 10–1 win. Drysdale hit two home runs to tie Don Newcombe's league record of seven for a pitcher.

Over ten games the Dodgers went 8–2 and pulled within two games of .500 at 59–61. They were still eleven games behind the Braves, with no realistic chance for a run at the pennant, but a third-place finish was within sight, a respectable conclusion to a forgettable season. The Reds, who had been easy marks for the Dodgers all year, arrived in Los Angeles for a four-game series and won three, starting the Dodgers' slide back to the depths. The Giants pounded them 14–2 on September 1 to drop the Dodgers to fifth place, and the next day they lost a doubleheader and fell another notch to sixth. On September 6, the Dodgers won their final game of the year against the Giants, 2–1. San Francisco had proved hospitable to the Giants, who climbed from sixth place in the Polo Grounds in 1957 to third in Seals Stadium.

The Dodgers concluded their work with the Phillies for 1958. Philadelphia, the only team with a worse record than Los Angeles, provided no relief, sweeping a four-game series. The Dodgers then split a pair with Pittsburgh and another with Milwaukee. A doubleheader loss to the Reds left them in sixth place, more than twenty games behind the Braves. All that remained were six games with Chicago and two with St. Louis, as all three teams battled for fifth place. The Dodgers finished their road schedule by taking two of three from the Cubs in Wrigley Field, then returned to the Coliseum to split a pair with St. Louis. On September 26, the Dodgers divided a doubleheader with the Cubs, winning the opener 6–3 and dropping the finale 2–1. The next day the season ended mercifully with the Cubs wresting sixth place by defeating the Dodgers 7–4.

The crowds, which had been so large and enthusiastic in April, had dwindled to 10- and 12,000 for the final home stand. The total attendance for the season was 1,845,268, in spite of the Dodgers' worst finish since 1944, when their team had been decimated by the military draft. The attendance fell short of the Braves' record, but surpassed the club mark established in 1947, Jackie Robinson's first year in Brooklyn. The average attendance had grown dramatically from 13,600 for the final year of Ebbets Field to 24,000 in the converted Olympic and football stadium. Although hampered by a poor facility and a worse team, the Dodgers were estimated to have made $4 million at the gate, even after deducting the visitors' cut.[43] Financially, the move was so far a spectacular success, justifying O'Malley's most optimistic hopes.

The performance on the field had some bright spots, as well. Several players rallied to post respectable records. Gil Hodges raised his average to .259, with twenty-two home runs. Charlie Neal matched that home run total and hit .254. Duke Snider, who struggled all season with a damaged knee, ended at a .312 average, with fifteen homers. Johnny Roseboro hit .271, with fourteen home runs, and Carl Furillo batted .290, with eighteen. Among the younger Dodgers, Don Zimmer tailed off in the second half, his average falling to .262, but he hit seventeen home runs, exceeding Reese's highest career mark. Dick Gray, who had started so well, hit no homers in the second half, and Gino Cimoli batted only .246.

The pitchers fared poorly. Don Drysdale improved to 12–13 and lowered his earned run average to 4.17. Johnny Podres, who had started well, finished at 13–15, with an e.r.a. of 3.72. Sandy Koufax labored through the second half and posted a record of 11–11 and 4.48. Stan Williams, whose early career showed such promise, ended the season at 9–7, with an e.r.a. of 4.01. Two of Brooklyn's stars were at the end of the line. Carl Erskine finished at 4–4 to end a distinguished career. Don Newcombe won seven games in 1958 and another nineteen after that, but all for Cincinnati and Cleveland.

The *Los Angeles Times*'s sportswriters, cheerleaders in April, were rather disgruntled at season's end. Frank Finch described the Dodgers final loss to the Cubs as "the story of the Dodgers' life in Los Angeles . . . a story of ineptitude and frustration . . . a story of a team that couldn't win the big one."[44] This critique was especially sharp because it focused on the team's character more than its ability. He wrote, "Time after time this sad season Walt Alston's erratic athletes failed to meet the challenge when just a little extra oomph would have improved their status."[45]

Paul Zimmerman, the *Times*'s sports editor, wrote a column on September 29 entitled, "L.A. Fans Entitled to Stronger Team," in which he complained, "We've had all the alibis—the move West, the short left-field and long right-field fences, the Chavez Ravine situation, injuries, etc., etc. It is high time action supplanted alibis."[46] Zimmerman noted that O'Malley had admitted an error early in the season: the club had continued to rely on the Brooklyn veterans who were past their prime rather than promote the young players who would compose the next Dodger team. Zimmerman dismissed this explanation: the youth movement contained only one or two players of sufficient quality to improve the Dodgers' fortunes. Nor did he anticipate winter deals with much enthusiasm.

The *Times*'s pessimism was not shared in the Dodgers' executive suite. O'Malley offered his own assessment of the season early in September, and his ebullient evaluation focused on future attendance records. "I am confident that if we could have had the same start as the Giants did this year we could have hit 3,000,000 at the Coliseum here. No club has ever done it. I feel certain that on the basis of our showing this year, Los Angeles is the one place in baseball where it can be done. When we have the team, we'll hit it!"[47] O'Malley exulted in the expected fall of the Brooklyn attendance record, and noted that the achievement came in spite of the team's poor performance and the lack of traditional amenities such as beer. His plans for 1959 included bringing in the right-field fence and scheduling more night games to solve the heat problem.

The contrast between the *Times*'s assessment of the season and O'Malley's points up some important characteristics about the Dodger organization. In other organizations, managers have been fired for winning only pennants and not the World Series. And the occasional bad season, inevitable for every team, has led other owners into impetuous moves that have weakened teams for years to come. But the Dodgers have consistently taken a different approach. They operate on the premise that they are in the entertainment business. Their competition is not only the Giants, Reds, and the Yankees but, perhaps more importantly, the movies and beaches of Southern California. To compete against these and other attractions, the Dodgers must field quality teams and on occasion win pennants; but they must also provide affordable entertainment in an attractive facility with adequate parking.

They have prospered greatly in a community where entertainment is a dominant industry, and their first season in Los Angeles established the organizational foundation of that later success. Faced with a crushingly disappointing season, the Dodgers did not panic: Walter Alston's contract was renewed for 1959 even before the season ended; player moves were kept to minimum; and no rebuke was uttered publicly from the owner's office.

The team was trounced regularly by the Giants, spent most of the season in last place, and collapsed into seventh place after reaching the first division in August. Yet, throughout the season, O'Malley kept the key issue in focus. No matter how the team performed on the field the most vital goal in the first year was to secure the Chavez Ravine agreement, despite the challenges which threatened the entire basis for the move West. Frank Finch complained that the team could not win the big game because they lacked the requisite effort at the key moment. The same could not be said for Walter O'Malley, who understood that the most important contest of the year was the referendum on June 3, and who, along with Mayor Poulson, approached that battle with a textbook example of political campaigning.

O'Malley's emergence as an active figure in the referendum fight represented a change in strategy. When the referendum was placed on the ballot in December 1957, O'Malley declared he would remain on the sidelines and not attempt to influence the voters.[48] This detachment was based on the assessment that 70 percent of the electorate favored the Chavez Ravine agreement. Then opponents of the deal began to win converts with their call for a better contract. Led by council members Pat McGee and John Holland, the "resourceful enemy," as Mayor Poulson characterized them, fashioned a strategy that appealed both to sports fans and tax payers. Poulson's later account of the controversy focuses on the role of Councilman McGee. "Handsome, boyish, and blue-eyed, he told the citizens that naturally he favored baseball. He merely wanted a more equitable contract. In truth, he was friendly with a group that wanted O'Malley in Chavez Ravine like it wanted smallpox."[49] The group to which Poulson refers was the Smith brothers, the owners of the San Diego Padres who had been adversely affected by the removal of Pacific Coast League franchises from Los Angeles and Hollywood.

The alarming news that only 52 percent of the electorate favored Proposition B forced Poulson and O'Malley to abandon their detachment. Poulson candidly admitted urging the Proposition's leaders to scotch the approach of "using carefully reasoned arguments to justify the contract."[50] Poulson's alternative strategy was a shrewd campaign equating the referendum with baseball in Los Angeles and tying baseball to mass entertainment. He devised a class-based argument linking opponents of Proposition B with an elite who would still enjoy the diversions that wealth and privilege could buy while the working-class would be deprived of major league baseball.

In a remarkable admission, Poulson later noted, "I suggested a scare campaign that would strike home with the low-income people who didn't belong to country clubs and social groups and who wanted big league baseball for entertainment. The referendum, we led them to believe, was unalterably a yes-or-no vote for baseball. By this time the Dodgers had started playing in the Coliseum, and the fans loved them. The prospect of losing them wasn't appealing."[51] While Poulson played his own brand of hardball, a citizens committee for "Yea on Baseball" promoted the referendum. Headed by comedian Joe E. Brown, the committee arranged periodic endorsements of the proposition by community leaders. At regular intervals, representatives of little league, labor unions, the medical community, and other groups uttered public support for the value of keeping the Dodgers in Los Angeles.

As the new season approached, the campaign over the referendum accelerated. At the same time another tactic to block the Chavez Ravine agreement emerged—suits filed in Los Angeles Superior Court to nullify the contract between the city and the Dodgers.[52] On April 18, the Dodgers played their first game in the Coliseum and Judge Kenneth Newell heard arguments against the Chavez Ravine pact. Councilwoman Wyman, only a week after giving birth to a daughter, attended the first major league game in Los Angeles, and John Holland, her foe, announced he had discovered a cemetery interested in the Chavez Ravine site.

The state government joined this political circus, the Assembly Interim Committee on Government Efficiency and Economy proclaiming it would hold hearings in mid-May to study the Dodgers' agreement with the city. These began on May 15, and included a jolt for the referendum's supporters when Karl Rundberg, a councilman, switched positions: he now opposed the Chavez Ravine agreement.[53] Rundberg had wavered about voting for the contract in 1957, relenting only after intense pressure from Councilwoman Wyman.[54] Rundberg explained he had changed his mind after listening to testimony that described the contract as an ambiguous document and he cited more pressing problems which the city ought to confront. Rundberg favored renegotiating the contract, a strategy that presumably would have taxed city resources even further. Nonetheless his remarks were applauded by the audience that had filled the committee room and obviously opposed the contract. Rundberg's announcement was countered somewhat by O'Malley's testimony later that day. Committee members began their questioning with queries about the team's inept performance. O'Malley had a ready answer. "Our boys came out here and instead of feeling that they knew what the future meant for them, they started to read a lot of stuff about political agitation, and they don't know to this day whether they are going to stay here or not."[55] He then discussed the impact of the problems of players whose families were still in the East (a problem that evidently didn't hamper the Giants' play). He also discussed the defects of the Coliseum, and concluded, "I think we have to recognize the fact that the players have not been at ease, and I think after this election on June 3rd when all of this dies down, one way or the other, I think they will come through and play up to their capacity."[56]

In a line of questioning remarkably similar to what had occurred in New York on numerous occasions, O'Malley was asked what would happen to the team if his stadium were not built. "Our intention is to stay in Los Angeles," he replied. "We sincerely hope that we have a new ball park in which to play. And of course, it is very nice to want to stay some place."[57] Among the Dodgers' resources transferred from Brooklyn to Los Angeles was Walter O'Malley's ability to evade any issue.

At the second and final day of committee hearings, the contract's opponents presented their arguments that the deal between the city and the Dodgers needed renegotiation. Pat McGee charged that the contract was illegal, but he avoided personal attacks on Poulson or the other supporters: "I say no one is the villain in this case. The villain is the contract."[58] A member of the assembly, Lester McMillan, defended the public officials in Los Angeles who had drafted the Chavez agreement. "We have a very good Mayor and City Council," McMillan stated. To which McGee replied, "Yes, sir, but a poor contract."[59]

McGee objected particularly to the Dodgers' offer of Wrigley Field for Chavez Ravine. Describing the Angels' old park as a "white elephant," McGee said, "The Dodgers don't want it. So they just threw it into the deal to make it look like a good deal." McGee further denied that he represented the interests of the San Diego Padres or any other group. "I'm not getting a cent for this. I did get two tickets to the Dodger baseball games, however."[60]

Mayor Poulson and Councilwoman Wyman were called to rebut the testimony of the contract's opponents. Poulson couched the argument as one in which his concern for the city as a whole was challenged by the narrow interest of individual council districts. "Baseball is going to do a great deal for all of Los Angeles," he remarked, "and I can't afford to represent just one district."[61] This theme was expressed in the May 18 edition of the *Los Angeles Times*, whose lead editorial reviewed Councilman Rundberg's position: "His argument suggested nothing more than the fact that the city has become fragmented. If any public enterprise is not directly profitable to a particular district, that district is perfectly proper in opposing it. God help the city, if that principle is always to be applied."[62]

Additional drama was interjected on May 22 when Warren Giles, the National League President, announced, "If the vote on the city's Chavez Ravine contract with the Dodgers is refuted by the citizens of Los Angeles, it will be my personal recommendation to our league that we take immediate steps to study ways and means of relocating the franchise in another city."[63] Giles noted that the location of franchises was controlled by the league, and that New York had been vacated only because of the compelling long-run advantages believed to exist in California. A vital component of those advantages was the Chavez Ravine contract. Giles implied that permission to play in the Coliseum was granted only because it had been assumed that construction on the new stadium already would have begun. Giles extended his threat. "The June 3 referendum is of great importance to the National League and, as you know, to Los Angeles. It is more than the Chavez Ravine issue. To me it will be an expression by the people of Los Angeles as to whether they want major league baseball."[64]

After decades of pursuing a major league franchise, the clear implication was that if Los Angeles rejected Proposition B it would not only lose the Dodgers but remove itself from further consideration by another major league club. Giles concluded with the observation that he hadn't anticipated these obstacles and that he couldn't believe the referendum would be defeated in light of community feeling as well as the exceptional attendance the Dodgers had enjoyed in the Coliseum. He expected the contract to be upheld, and that the Dodgers would fulfill the league's wishes by securing a permanent home in Los Angeles.

Giles's comments triggered a predictable reaction. Councilman Holland denounced the statement as "the last desperate threat of a frightened group of greedy men." McGee added that it was an "insult to the intelligence of the people of the city of Los Angeles."[65] The following day Giles clarified his statement: "All I am doing is stating the facts. I am not presumptuous enough to attempt to indicate how the citizens of Los Angeles should vote."[66] He denied he was threatening Los Angeles or issuing an ultimatum; he was simply pointing out that the Dodgers' move had been predicated on the very contract now being challenged through the referendum. If the contract were nullified, then the basis for the transfer would be removed, and the team could seek another city.

Mayor Poulson agreed with Giles's assessment: "Los Angeles would be the laughingstock of the nation if we went back on our word."[67] Poulson's observations were offered from a unique perspective—he had orchestrated Giles's provocative remarks. In his memoirs, Poulson refers to this episode as "another instance where I have to pat myself on the back."[68] Describing the motivation for getting the league president into the act, Poulson writes: "I personally called Warren Giles . . . on the phone in Cincinnati, to discuss the campaign with him. I convinced him that he should make a terse statement that this issue was definitely whether we were going to have baseball or not and the people's vote would decide it at this election. This cleared the air of confusion and certainly made the opposition angry."[69]

The opponents' distress was understandable; their campaign was based on the position that they favored keeping the Dodgers in Los Angeles but on the basis of a better contract than the one negotiated in 1957. Poulson and Giles had blunted that argument by appearing to construct the issue as one of a yes-or-no vote on the Dodgers and on baseball itself. Walter O'Malley was thus in a position to take the high road. In his first public response to Giles's announcement, O'Malley stated, "The National League does have the right to move the franchise, but I shall fight any such attempt with all my strength. The players and our staff want to stay in Los Angeles. We like the location, the weather, the fans and the attendance records. We plan to be in Los Angeles permanently. I pledge myself to try to keep major league baseball here."[70]

O'Malley again expressed his confidence that the referendum would be approved; but for the first time he attacked the opponents of the contract. John Smith, owner of the San Diego Padres, was dismissed as a "minor league man who doesn't want major league baseball in Southern California." Of the two council members who had opposed the agreement all along, O'Malley said, "One doesn't understand what it's all about and the other merely has political ambitions."[71] In a statement to press, O'Malley reviewed some of the issues surrounding the Dodgers' move to Los Angeles and reiterated his arguments about his desire to build a stadium with private capital. He rejected any speculation about renegotiating the Chavez Ravine contract in the event that the referendum failed.

On May 27, John Holland created a brief furor by revealing a press release drafted in 1957 by Roger Arneburgh, the city attorney, and cosigned by the county counsel, Harold Kennedy.[72] The release stated the terms of an agreement between Los Angeles and the Dodgers to bring the team from Brooklyn. In a corner of the release was a handwritten message: "Vetoed in toto, WFO'M." The release was dated May 2 but was never issued, apparently because of O'Malley's objections. A slight modification of the statement was issued on May 6 but included only Kennedy's signature.

Holland's revelation, staged carefully for the benefit of television cameras gave him the opportunity to denounce O'Malley's involvement in city government: "This participation in the intimate operation of our city and county government by a resident of Brooklyn while visiting Los Angeles to seek concessions from the city, is extraordinary and certainly the cause for reflection. I think the matter deserves the most careful consideration of every member of this Council, as well as every voter in the city."[73] Arneburgh offered an immediate rebuttal, dismissing the charges as "ridiculous." He explained that the release and O'Malley's veto were old news, and that neither represented any control held by the Dodger president over the internal workings of city government. O'Malley's objections, Arneburgh explained, were to the press release's specific reference to the Dodgers. The subsequent changes referred to "bringing major league baseball to Los Angeles."[74] In fact, the original purpose of the Arneburgh memorandum had been to nullify the overly generous terms extended to O'Malley during the spring training meetings in 1957. O'Malley himself described the release as "a red-herring and an old one." He stated that he had not approved the press release of May 2 "for the reason that on that date we were considering ways and means of bringing major league baseball to Los Angeles and not specifically the Brooklyn Dodgers, as the Dodger ownership at that time was not ready to give up efforts being made in New York to keep the franchise there."[75]

This controversy set the tone for the final days before the referendum vote. On May 28, Pat McGee proposed a new plan for the Dodgers in Chavez Ravine. He favored offering the Dodgers just enough land for a stadium while the city retained the rest of the site. This proposal brought a sharp rejoinder from another council member, James Corman, who asked McGee, "Has this proposal been approved by J. A. Smith?" McGee snapped, "That's a damn dirty snide remark."[76] McGee added that Smith had nothing to do with the offer, and the two members reportedly "glared at each other across intervening Council seats."[77] At that point the council president, John Gibson, called for more decorum.

McGee made another pass at Warren Giles, pointing out that the National League constitution required unanimous approval for a franchise relocation, so that O'Malley could veto any attempt by the league to force the Dodgers out of Los Angeles. In fact, league bylaws had been modified to permit a shift with a three-fourths vote, a point that O'Malley was happy to make later in the day in a trenchant fashion: "This misleading statement by Councilman McGee is typical of the tactics of the opponents of Proposition B. The Dodger vote alone could not keep the franchise in Los Angeles as Councilman McGee improperly stated."[78] The Dodgers' president did not speculate about the likelihood of the league expelling him from Los Angeles against his will.

In that same council meeting, Charles Navarro, the Finance Committee chairman, addressed the claims of the referendum's opponents that Chavez Ravine was worth between $18 and $30 million. He noted that when the city purchased 169 acres from the Federal Housing Authority in the early 1950s the land had been appraised at $1,279,000. Navarro released other information showing that the value of land in Chavez Ravine had declined from assessments made in 1925. "These are realistic figures anyone can understand and sharply point up the incongruity of these $30,000,000 dreams," Navarro concluded.[79] Part of the difference in these assessments of the land's worth may have been based on comparisons between its current and potential value. Once properly graded, Chavez Ravine represented the largest undeveloped site in central Los Angeles. Grading, however, required significant investments and could be undertaken only for a project promising substantial commercial returns. As Poulson's memoirs suggest, no firm had been willing to risk such a venture until the Dodgers arrived.

On Sunday, June 1, the *Los Angeles Times* published its final editorial in support of the referendum. It framed the issue in the context of what kind of city Los Angeles might become. "Do you, a citizen-voter, want Los Angeles to be a great city, with common interests and the civic unity which gives a great city character; or are you content to let it continue its degeneration into a geographical bundle of self-centered sections each fighting with the others for the lion's share of the revenues and improvements that belong to all?"[80] Hyperbole aside, the editorial makes a compelling point. In New York, the jealousies of the various boroughs, reflected by the Board of Estimate, precluded any accommodation of O'Malley's interest in a new stadium and figured in Brooklyn's loss of the Dodgers. Although Los Angeles was not organized along the lines of a federated system, the size of the city deterred perceptions of common interest. As the *Times* concluded, "It is enough simply to remember that a vote for Proposition B will help to restore the communion of citizens in Los Angeles that has been dissolving so alarmingly."[81]

Later that Sunday, in classic Southern California fashion, a telethon paraded entertainers and celebrities who lobbied for approval of the referendum. George Burns, Groucho Marx, and Jack Benny led the notable cadre promoting the Dodgers' cause. One actor defended the Chavez Ravine agreement with great vigor: "I have always believed there are two sides to every question, but in this case they are the good side and the bad side. Chavez Ravine has been sitting there in the heart of Los Angeles for years and nothing was done with it. Now that a baseball team is to have it, it's worth a lot of money we are told. Sure Walter O'Malley got a good deal when he was offered Chavez Ravine as a site for his ball park. Any deal to be good must be fair to both sides, not just to one."[82] Ronald Reagan went on to decry the argument against Proposition B as "One of the most dishonest documents I ever read in my life. . . . For years we have been watching golf courses and other recreation areas destroyed to make room for subdivisions and factories. Where is a baseball stadium to go, in the suburbs, away from the freeways?"[83] After five hours the telethon concluded and the performers adjourned to the airport to greet the Dodgers on their return from Chicago. The *Times* reported that switchboards received hundreds of telephone calls in favor of the Proposition.

The final event of the campaign occurred on Monday when Walter O'Malley "debated" John Smith on a local television station. The session was less a debate than a recapitulation of positions, with the *Times*'s coverage neglecting to provide much ink for the opponents' arguments. O'Malley mentioned that he was weary of the tumult surrounding the agreement between the Dodgers and the city and that he was prepared to abide by the voters' decision. He added that the team's poor performance was perhaps a blessing in disguise since "it might be in the script for the Los Angeles to develop a Dodger team of its own from fresh young players."[84]

On June 3, the voters at last were able to render a decision. On June 5, the *Times* reported that returns from 2,350 of 4,519 precincts indicated 176,605 votes cast in favor of Proposition B and 159,028 in opposition.[85] The margin was considered decisive by both sides. Mayor Poulson voiced his approval of the vote and Earle Baker, a councilman who had voted against the agreement in September of 1957, said he accepted the vote and would no longer oppose development of a stadium in Chavez Ravine.[86]

The *Times* estimated a final margin of victory of 30,000 votes, but it turned out to be only 24,293 of 666,577 votes. The narrowness of the outcome reflected in part the voters' concern about their fiscal interests and whether those interests could be trusted to politicians. More importantly, the attraction of major league baseball and the status it conferred on the city overrode financial concerns. Walter O'Malley expressed pleasure at the margin of victory: "I don't regard the outcome as close, but rather as a very significant margin. Why? Few Presidents get elected with a greater majority than the 52½% indicated by the latest count. A poll we had privately conducted indicated that the voters were 67% favorably inclined at the start of the referendum, but a sampling after the club slipped into eighth place indicated a drop to 37%. Last week the poll prediction was 51.3%, which just about hit the nail on the head."[87] O'Malley was grateful to the city. "We sincerely appreciate the support of those who voted 'Yes' on Proposition B, and we will be zealous in our efforts to gain the support of those who voted 'No' by continuing our practice of living up to all of our responsibilities under the terms of the contract we have signed with your elected officials."[88]

Construction of the stadium was expected to begin in the immediate future. The grading of the hilly terrain could begin in a month, according to estimates O'Malley himself agreed, with the caveat of a man keenly aware of the constraints of government on commercial enterprises. "It is entirely possible," he remarked, "that we can get going by July 5, if there are no road blocks such as problems of proper clearance or delays due to litigations."[89] His qualification was well founded. On June 6, superior court judge Kenneth Newell issued a preliminary injunction in a taxpayers' suit blocking transfer of Chavez Ravine from the city to the Dodgers. Before the wheels of justice permitted the first spade of earth to be turned in Chavez Ravine, the Dodgers would rise from the horrendous season of 1958 to pursue a world championship.

8

Due Process and Another Championship

At the same time public attention focused on the referendum on the city's pact with the Dodgers, taxpayer suits filed by Julius Reuben and Louis Kirschbaum sought to block the transfer of the Chavez Ravine property.[1] These suits need to be seen as a distinct challenge to Walter O'Malley's objectives because the judicial branch exists in part to stop illegal government action which even a majority of the population may favor. In contrast to the freewheeling political process of bargaining, electioneering, and celebrity telethons, the judicial process is characterized by rigorous and minute analysis of specific phrases and clauses in pertinent documents.

Whether a baseball team and a privately owned stadium qualified as a public purpose under the meaning of the Federal Housing Act of 1949 had been disputed by elected officials in New York and Los Angeles for several years. That issue would now be determined in a judicial forum.

Important differences distinguished Chavez Ravine from Atlantic-Flatbush, and those differences could be part of the court's determination. The significance of these cases was barely noted by the *Los Angeles Times* in its extensive coverage of the successful passage of Proposition B. A brief column noted that three suits had been filed challenging the contract and that a temporary restraining order had been issued on April 28 blocking execution of the contract pending a hearing on a request for a preliminary injunction.[2] Granted just three days after Proposition B passed, the injunction enjoined city officials from violating any part of the public purpose restriction included in the deed that conveyed Chavez Ravine from the City Housing Authority (a state agency, despite its name) to Los Angeles. Because the precise meaning

162

of the public purpose restriction remained to be determined, the effect of the injunction was not terribly significant. In anticipation of future litigation, Mayor Poulson had prompted a rewording of the restriction but no court had determined that a privately owned baseball stadium was definitely beyond the scope intended by the troublesome phrase. Julius Reuben, an attorney who represented himself in the proceedings, predicted the litigation would ultimately be reviewed by the United State Supreme Court.[3]

The hearing, which first addressed the substantive issues surrounding the contract, began on June 20, 1958, in Los Angeles County Superior Court. The presiding judge was Arnold Praeger. The case joined the suits of Reuben and Kirshbaum, who was represented by Phill Silver. Silver's arguments implicitly assumed that the Dodgers did not constitute a legitimate public purpose. He focused his attack on the city's request, complied with by the Housing Authority, to eliminate the public purpose restriction from the deed of transfer.[4] Not only was that removal sought by Poulson and agreed to by the Housing Authority; the ordinance passed by the city council on October 10, 1957, and the basis for O'Malley's announcement that the Dodgers would move to Los Angeles, included the clause that the city "agrees to use its best efforts to eliminate or modify a public purpose restriction on lands purchased by the City subject to such a restriction."

The brief Silver submitted to the court repeatedly charged that none of the city officers nor the Housing Authority had the power to make such a change in the Chavez Ravine deed.[5] Silver made seventeen arguments that challenged in detail the contract between the city and the Dodgers. Some addressed the manner in which the deal was reached, such as the absence of sealed bids for the property by competing parties; another argument alleged the contract was void because it lacked mutuality, a reference to the purported disparity between the value of Chavez Ravine and the worth of Wrigley Field, which O'Malley had exchanged for the new stadium site. In reply, attorneys for the city and for the Dodgers contended that the mayor and the city council had acted within their appropriate discretion when they agreed to the contract.[6] They objected frequently that questions and arguments made by Ruben and Silver were immaterial since they did not concern the narrow points of law being contested.

This particular controversy goes to the heart of the judicial function. In the legislative and executive branches of government representatives are elected for the purpose of establishing policies that promote the good of the community at the present time and in the future. The function of the judicial branch is to resolve disputes over laws and facts that have already occurred. Judge Praeger's responsibility was limited to determining whether the contract had been reached in conformity with established legal rights and responsibilities.

When Walter O'Malley was pursuing new quarters for the Dodgers in New York, the elected and appointed representatives of that community determined it was not in the public interest to accommodate O'Malley's request. Some, such as George McLaughlin, also speculated that courts would not uphold the interpretation of the public purpose clause of the Federal Housing Act even if land were transferred to the Dodgers under that authority. Public officials in Los Angeles understood the meaning of that law and the interests of their constituents differently. Judge Praeger was faced with the task of determining if the method and substance of the contract conformed to the delegation of power under both the charter of the City of Los Angeles as well as the Housing Act. The exquisitely complex nature of the judicial process was evident as Mayor Poulson took the stand to explain his role in forging the contract that had lured the Dodgers from Brooklyn.

For all the light which Poulson might have shed on the many disputed points, he was able to answer only one question, even though he was on the stand for an hour.[7] Again and again, questions from Silver were challenged by the attorneys for the city, and those objections were sustained by Judge Praeger. The one query to which Poulson replied concerned his letter to the Housing Authority seeking modification of the language in the original deed to remove the public purpose provision. Silver asked, "Was it your intent and purpose that the Housing Authority should issue a new deed to extinguish the public purpose restriction?" Poulson answered, "I presume I would have to say yes."[8] Silver followed by asking if Poulson still intended to have the Housing Authority make that change at the time the taxpayers suits were filed. Praeger interjected that "the city bound itself by contract and ordinance to try to bring this about. It doesn't make any difference what the Mayor did."[9] Silver protested that he needed to show that the mayor was trying to use the Housing Authority to make the change, and Praeger asked if Silver was alleging that the mayor lacked that right. Silver replied that indeed he considered such a request illegal. Praeger ruled that the line of questioning was immaterial. He expanded, "Let's get down to some basic principles. The City Council enacted an ordinance. The court can't supplant its ideas of what should be done."[10]

The virtual futility of Poulson's appearance demonstrates the distinctness of judicial challenges to the Chavez Ravine contract. Where Proposition B had simply asked voters if they thought their elected representatives had acted in the public interest, the taxpayer suits fragmented the essential issue of the validity of the contract into a host of subordinate legal questions that could be addressed only if they conformed to the narrow strictures of these particular judicial proceedings.

An implication of this sort of judicial challenge is that its issues are almost hopelessly inaccessible to anyone but a lawyer. A second implication—of devastating consequence—is that no one can participate in this exercise without the funds to hire skilled attorneys. Even the successful party incurs significant costs in expenditures for counsel. In this case there was the added potential cost of delay in the construction of a new facility.

On the day of Poulson's testimony, the Dodgers attempted to diffuse one controversial issue. O'Malley sent a resolution from the team's board of directors to the city council renouncing all claims to oil rights in Chavez Ravine.[11] The contract had provided that any oil revenues would go to the city but that they would be used in part for funding the Dodgers' youth program. This provision has contributed to a notion, which still persists, that the city gave oil rights to the Dodgers, an especially notorious proposition after the oil shocks of the 1970s.[12]

Councilman Holland tried to use O'Malley's resolution as an opening through which to force a renegotiation of the entire contract. Holland contended that discussions of the proposed renegotiations might ensue, "If by any chance of the super power of money this contract is found legal."[13] He denied he was casting aspersions on the integrity of Judge Praeger's court; he wished only to emphasize that the Dodgers' money could provide them with superior legal talent than their opponents could muster. In any event, when the Dodgers' counsel attempted to introduce the abandoned oil rights as a point for consideration in the taxpayers' suits, Judge Praeger ruled it was inadmissible as evidence because it, too, was immaterial.

On the next day of testimony, Judge Praeger directly addressed the public purpose argument. He questioned whether the public purpose restriction had any real significance. Praeger interrupted one of Phill Silver's arguments to remark, "If I understand the law, any property conveyed to a government is for public use. The city can't acquire property for private purposes. So I can't see that the words have any real significance."[14]

The judge then noted another restriction in the deed that reflected the hostility to public housing which had originally stalled the land acquisition. The restriction prohibited the use of Chavez Ravine for a residential subdivision for a period of twenty-nine years. Praeger concluded that "This carries the implication that the city can sell this property, provided it is not used for that specific purpose (residential sub-division). That negates the idea that the land was to be used perpetually for public purposes."[15] Praeger asked if the public purpose restriction, given this interpretation, retained any meaning. Silver contended that it did, but the judge's reading suggested that the city might acquire land only for a public purpose, but that at some later time the land could be sold to private parties for their own use. As muddled as that distinction might be, it doesn't even reach the issue whether the use of Chavez Ravine for a privately owned baseball stadium constituted a public purpose.

Robert Moses had chided O'Malley for thinking that the Dodgers' interest in a new stadium conformed with the public purpose provisions of the Housing Act, but Los Angeles officials had pursued the team with great earnestness precisely because they saw a major league team in the city as a means of enhancing the status and reputation of the largest city in the western United States. Having failed to find an alternative for Chavez Ravine, Los Angeles was in a position to argue that the Housing Act could indirectly be used for a private baseball stadium.

Silver concluded his arguments by continuing to press the theme that the city could not transfer a public asset for the benefit of a private business. Praeger complied with Silver's request for a temporary re-straining order to prevent certification of the June referendum on Proposition B until the judicial process had determined if the contract was in fact legal.

On the following day of the proceedings, the city's attorneys introduced their arguments, and Praeger challenged many of their contentions. One point of concern was the contract's provision that forty acres of Chavez Ravine be used by the Dodgers to provide recreational facilities to the public for twenty years. At the end of that time, the land could be used by the Dodgers for whatever purpose they wished. The contract required the team to spend no more than $500,000 on the recreational facilities. Praeger was curious about the consequences if O'Malley chose to spend significantly less than the limit.

Bourke Jones, an assistant city attorney, replied, "If they did that it would be solely the fault of the city, because they must mutually agree on the facilities to be built."[16] Jones suggested that a court of equity could resolve any dispute between the city and the Dodgers over what kind of facilities should be provided. Jones continued his presentation by arguing that the private benefits which befell the Dodgers from the Chavez Ravine contract were incidental to the public purposes being served, as indicated in the city council's resolution. He then alleged that any dispute over oil rights was moot since not a drop of oil had even been discovered either in Chavez Ravine or Wrigley Field.

Julius Reuben also introduced his remarks that day. One of his major points was immediately dismissed by Judge Praeger. Reuben intended to argue that the city council had been guilty of gross negligence for failing to obtain a suitable appraisal of the value of Chavez Ravine. His argument was interrupted by counsel for the Dodgers, who pointed out that in a pretrial hearing Reuben had contended that property values were not an issue in the case. After checking the records of the proceeding, Judge Praeger concurred with the objection and, to Reuben's chagrin, he barred the attorney from continuing that line of argument.

Reuben then took up points that had largely been expressed earlier by Silver. He contended that the city council had abused its discretion by transferring land without the deed restrictions. He also raised points about oil rights and the recreational facilities. In his summation Reuben posed the fundamental issue: "The question is whether the contract is a gift of public property under the law. The whole question hinges on whether the ball club is a private or public purpose."[17]

On June 25, the day Judge Praeger took the case under submission, a more discreet issue was explored. A provision of the contract required the city to vacate public streets in order to make room for the stadium. The judge interrupted the Dodgers' attorney to ask, "Is it not possible that the agreement to vacate streets does not involve the public interest but benefits a private interest?"[18] The attorney, Pierce Works, replied that there might be an "incidental" benefit to the Dodgers, but that the contract permitted the closing of streets only if it has been determined that the roads were not needed for public convenience and necessity. Praeger asked how the contract differed from one in another case that had been found invalid because a street had been closed to benefit another corporation. "In this contract," Works answered, "the city does not say it will close the streets whether they are needed or not. The City Council still must make a determination that the public convenience and necessity warrants the closing."[19]

Praeger informed the attorneys that he expected to reach a decision in a week to ten days. The litigation had progressed rather expeditiously because the disputes concerned interpretations of only several clauses in the Chavez Ravine contract and precluded the need for the parade of witnesses that characterized trials in which fundamental disagreements over facts are at issue. The Reuben and Kirshbaum suits turned more on the arguments of counsel, who were prodded along by a judge who found all extraneous considerations inadmissable.

The referendum took Walter O'Malley by surprise; but he must have anticipated a judicial challenge to his contract with Los Angeles. In New York he had expected such suits in the event city officials complied with his requests there. Nevertheless, litigation meant further delay, expense, and uncertainty in his pursuit of a permanent new home for the Dodgers. On July 14, 1958, Judge Praeger issued the stunning ruling that the Chavez Ravine contract was invalid. He wrote, in part, "This is an illegal delegation of the duty of the City Council, an abdication of its public trust and a manifest gross abuse of discretion.[20] The decision nullified the agreement framed in 1957 which brought the Dodgers west from Brooklyn and which had survived the arduous battle over Proposition B. The political exercises of elected and appointed officials and of the electorate fell before the ruling of Judge Praeger that the contract did not conform with legal requirements concerning the delegation of power. He specifically objected to the contract's provision to give the Dodgers half the oil revenues, even though the team originally was required to use the money for its youth programs, and despite O'Malley's relinquishment of all claims to those revenues, and despite the fact that no oil had ever been found at any of the sites in question.

Praeger objected further to the clause requiring the city to abandon city streets for the stadium, even though the city retained the right to determine if such action conformed to the public purpose clause. Public funds, in Praeger's judgment, could not be used to obtain property "to be transferred to a private corporation for the operation of a private business."[21] With that remark Praeger answered the fundamental question that had surfaced when O'Malley first approached Robert Moses about providing land at Atlantic-Flatbush. The judge concluded that, in spite of the benefits which would befall the city through tax revenues, youth programs, and stimulus to other local businesses, and the importance to the town that it achieve "major league" status, the primary beneficiary of the transfer remained a privately owned baseball team.

Praeger's other objections included the contract's provision that the city spend up to $2 million to put the property in suitable condition to transfer to the Dodgers with the determination of how that money was to be spent left to the team. "The manner of expenditure of public money," the judge ruled, "is in the nature of a public trust and cannot be delegated to others. The City Council has no right or power to give a private organization carte blanche with respect to the spending of public money."[22] About the public purpose issue, Praeger ruled, "If the city can use public funds to acquire property for the purpose of conveying it to a ball club for a ball park to be operated as private enterprise for private profit, then the city may use public funds for the purpose of selling it for use for a private bowling alley, a private golf course, a steel mill, a hotel or any other private purpose."[23]

The decision took no note of the efforts Mayor Poulson had made to find a buyer for the property. Public housing had been rejected as a use and was now even precluded by law. The prospect for this sizable acreage adjacent to downtown Los Angeles was that it would remain undeveloped except for the community sparsely settled among its hills and canyons. Even today Roger Arneburgh finds no merit in Judge Praeger's decision, declaring it to be "flat wrong."[24]

After extended conversations with his attorneys, Walter O'Malley put the best face on the reversal. He stated his belief that the decision would be overturned on appeal and that the contract would ultimately be sustained. He regretted only the further delay. "I remain an optimist," he remarked at a press conference. "I think things ultimately will be resolved in favor of our contract with the city. I had hoped for a favorable ruling now, so that the injunction might be dissolved and preparations made to start the grading and construction. But this is just another hurdle which we will have to take in stride. What hurts is the delay. Our timetable is out the window and I'm afraid San Francisco will have its new stadium first."[25]

O'Malley might have been comforted if he had known what a miserable facility the Giants would find themselves in, but he reiterated his unwillingness either to renegotiate the original contract or to look elsewhere in Los Angeles for a stadium site. "We have had many other parcels of land offered us, but we have never inspected any of them. I want to make it very clear, moreover, that we have not gone shopping for any alternate site."[26] He rejected again the use of either Wrigley Field or the Coliseum as a permanent facility: "We came to California in the first place because we felt it was a fine country and because we wanted to build a new, modern stadium. Chavez fits in perfectly with that plan and we are not abandoning the program."[27] The record crowds using the Coliseum were no inducement to stay there. For the 1959 season the right-field fence would be brought closer to the field, but nothing could be done to make the left-field barrier anything but a curiosity. Judge Praeger's decision was problematic for another reason. The contract between the club and the Coliseum Commission was due to expire after the 1959 season. Another interim agreement would have to be arranged until some permanent stadium was finally constructed somewhere.

O'Malley was less optimistic about the reaction of other National

League owners to further delay in finding a permanent home: "Frankly, it poses a problem. President Warren Giles has been very outspoken about our troubles, and I know the other clubs are much concerned whether we will have the new stadium ready for 1960." [28] He neglected to mention that Giles's outspokenness had been orchestrated by Mayor Poulson to benefit O'Malley during the referendum controversy. The tactic yielded fresh dividends now, as the Dodgers' president raised the possibility that other National League owners might force the Los Angeles franchise to move to more hospitable quarters.

At the time of Judge Praegers' ruling, the Cincinnati Reds were in town and their general manager and president were both asked to comment on the Dodgers' plight. Gabe Paul, the club's general manager, conveyed the realities of the relationships among the league's owners when he said "As for the possibility of the Dodgers being forcibly taken out of Los Angeles by the National League, it is technically possible but never would be done against Mr. O'Malley's wishes. It would be presumptuous to tell him what to do with his property." [29] Had Paul made such a statement after Giles' announcement in May, the outcome of Proposition B might have been different. In any event, Paul simply said aloud what everyone familiar with the league knew anyway.

Powell Crosley, the Reds' owner, even had a kind word for the Coliseum. "Despite what happened to us this trip [the Dodgers swept the four-game engagement] we have no complaints about the Coliseum. Nor have I heard any lately." [30]

One reporter mentioned a prospect that must have been chilling to O'Malley. Since the court had ruled the contract in violation of the public purpose restriction, then what about O'Malley's willingness to rent a municipal stadium constructed at the site by the city? [31] Poulson had written that O'Malley first sought such an agreement when discussions about the franchise shift began during the spring of 1957. The experience of the Brooklyn Sports Center Authority might have reinforced O'Malley's devotion to private enterprise. In any case, O'Malley dismissed the reporter's question as hypothetical.

The issue at hand was Judge Praeger's ruling. The city favored an appeal, which could have been blocked by a vote of the city council. A majority of the members, however, indicated that they would be guided by the advice of Roger Arneburgh, the city attorney. Both Poulson and Arneburgh were out of town on the day of Praeger's ruling, but after a day of deliberation they announced that the city indeed would file an appeal. Phill Silver encouraged the Dodgers and the city to waive their right to appeal in order to begin negotiations on a new contract, an offer rejected as insincere. "Despite all declarations to the contrary," said one member of the council majority, "the whole effort of those opposed to the Dodger contract all along has obviously been to exclude the baseball team from Chavez Ravine altogether and I take an extremely dim view of any offers now to renegotiate the contract."[32]

An appeal of Judge Praeger's decision could take an inordinate amount of time which complicated matters considerably given the pressures to vacate the Coliseum after the 1959 season. In September 1958, City Attorney Arneburgh voiced his concern about the delays of litigation and announced his strategy to expedite the process: "Unless we can get a writ of prohibition, I don't think there's a chance we can get this thing cleared through the courts before 1960, and even with that break, I estimate it will take another year and a half to grade and build the location."[33]

The writ of prohibition is a device to stop the proceeding of a court when the court's juridication over an issue has been challenged and there is no other quick remedy. Arneburgh and the Dodgers filed this writ with the California State Supreme Court, the state's highest court, in the hope that it would assume jurisdiction and review the entire controversy. The outcome of such a review could not be predicted, but Arneburgh preferred such action because the more customary appeals process could consume more than a year. The *Times* reported that if litigation prevented the Dodgers from moving into a new stadium in 1961, they retained the option of staying in the Coliseum, according to a paragraph in their rental agreement with the Coliseum Commission, which read, "Provision for 1960: In the event the Dodgers desire to use the Coliseum for all or part of the 1960 baseball season, they shall pay as rent thereof the same amount provided for the nine games for the year 1958–59, namely, 10% of the gross receipts and the commission to keep all of the receipts from the concessions."[34] The significantly higher rate which the team paid for the first few games of 1958 and 1959 would thus extend through the entire 1960 season, costing the Dodgers a much greater sum than they had been paying. Moreover, renegotiation of the terms of that option was unlikely since a reduction of the stipulated rent could be interpreted as an illegal gift of public funds.

The need for further judicial proceedings became evident later in September when the Los Angeles County Board of Supervisors decided to reallocate funds earmarked for highways into Chavez Ravine for the construction of roads in the five supervisory districts. The supervisors tried to allay concern that they were turning their backs on the Chavez Ravine project, even as $1,350,000 was divided among their districts. This diversion of funds had been proposed by Supervisor Frank Bonelli, who declared that the transfer "does not indicate lack of faith that the Dodgers will ultimately have a ball park," rather that "they may be in the courts for years and meanwhile we need the roads in other sections of the county."[35] The original agreement between Los Angeles and the Dodgers hinged on the cooperation of several units of local government. As those units saw their commitments suspended by a court, pressure built to shift attention to community needs. By the time the Dodgers were vindicated in the courts other officials might conceivably have replaced the group that had worked so diligently to attract the team. Norris Poulson, who was defeated for reelection in 1961 by Sam Yorty, blamed the controversy over the Dodgers for his loss, and Rosalind Wyman believes that her identification with the Dodgers' contract contributed to her later defeat.[36]

While the legal machinations continued, the Dodgers concluded their inaugural season with their ominous seventh-place finish. The winter was expected to produce both some revamping of the team and a state supreme court ruling on the petition for a writ of prohibition. On January 13, 1959, the court unanimously sustained the original contract between the city and the Dodgers and granted the writ of prohibition, preventing Judge Praeger from taking further action to block or delay the construction of the stadium.[37]

The court ruled that in issuing the injunction to void the contract Judge Praeger has exceeded his authority: "A court acts in excess of its jurisdiction if it attempts to enjoin the enactment or enforcement of a valid public statute or ordinance."[38] The opinion declared further that a writ of prohibition should not be issued where the propriety of the court's action turns on controversies about facts, but that in the Dodgers' case "none of the persons attacking the contract has pointed to any evidence which would present a substantial question of fact upon a material issue."[39] The court considered some of the contract's provisions, pausing to comment on the issue of oil rights, which had been crucial to Judge Praeger: "The ball club had given formal notice to the city that the club has waived and relinquished all rights conferred upon it under these provisions and has assigned to the city all such rights. The provisions being solely for the benefit of the club, it could properly waiver them, and there is no longer any reason to treat them as parts of the contract."[40] To the challenge that the contract was void because it left to future determination provisions about recreational facilities, a site for oil drilling in Chavez Ravine, and possible rental of Wrigley Field by the Dodgers, the court found that these were "matters which should not be treated as rendering the contract void."[41] If the city and the Dodgers could not reach agreement on these issues, a court could make a reasonable judgment for them within the provisions of the contract.

The central issue—public purpose restriction—was also found to be satisfied: "In considering whether the contract made by the city has a proper purpose, we must view the contract as a whole, and the fact that some of the provisions may be of benefit only to the baseball club is immaterial, provided the city receives benefits which serve legitimate public purposes."[42] The court decreed that the transfer of Wrigley Field, along with the twenty years of recreational facilities the Dodgers would provide at Chavez Ravine, "are obviously for proper public purposes."[43] These direct benefits erased the need to determine the significance of indirect benefits such as the city's having a major league baseball team and the return of Chavez Ravine to the tax rolls.

The court further rejected arguments that the city had failed to determine that the land at Chavez Ravine was no longer needed and that the city could not agree to purchase additional land for sale to a private corporation: "While a city may not purchase land when no public use or purpose is involved, in the present case the additional land is to be acquired for a proper public purpose, namely, that of furnishing the type of contract consideration which enables the city to enter into a bargain which it deems advantageous."[44]

The city's promised effort to remove the public purpose restriction was also found legitimate: "In view of the declaration by the council that the city no longer requires the use of the property, the city's promise to use its best efforts to have the deed restriction removed cannot be said to constitute an illegal agreement binding the city council to exercise its judgment in the future. Any possibility that the Housing Authority may have a right to complain of a termination of the public use upon a sale by the city will disappear if the city is able to obtain the Housing Authority's consent to eliminate the restriction."[45]

The court accepted the the provision that the city would spend up to $2 million to improve the property before transferring it in a manner to be determined by the Dodgers. The court found the arrangement beneficial: "The city could have paid the money to the club outright, as part of the contract consideration, without retaining any control over the manner or purpose for which the money should be spent, and the arrangement by which the city will keep possession of the money until it is spent and, in addition, retain some control of how it is spent, is an advantage to the city which does not militate against the validity of the contract."[46]

The city's agreement to vacate city streets, a point that troubled Judge

Praeger, was readily accepted by the high court: "It is an elementary rule that contracts must be construed, whenever possible, so as to make them valid rather than invalid, and the provision, when read as a whole, may be construed as not imposing on the city any improper obligation. The clear implication is that the agreement will apply only to such streets as the city may subsequently determine to be no longer needed for street purposes. The agreement does not purport to control or govern the exercise of the power to determine what streets may no longer be necessary but to provide that, when the determination is made, the city will commence the required statutory procedure."[47] After some further consideration of the street issue, the court reached its majestic conclusion: "Let the writ issue as prayed."[48]

The predictable reactions included Mayor Poulson's statement: "I am glad that the Supreme Court was able to see the forest instead of looking at the small trees, as apparently was the case in the lower court."[49] "One World Series in Los Angeles," Poulson added, "and every cent that Los Angeles has invested in this project will be repaid many times over. Progress must not be stopped in Los Angeles."[50] Walter O'Malley expressed his pleasure at the outcome and his lack of surprise that the contract had been sustained. With customary optimism, he announced that work on the stadium would begin within days and that the facility would be ready for the 1960 season. At this point the plan was to open with a limited capacity of thirty-two thousand and then expand to a final seating for fifty-two thousand. O'Malley admitted he had been surprised by the large crowds in the Coliseum but that he didn't intend to increase the new stadium beyond its original design: "It's an engineers' rule if you go to over 52,000 you sacrifice both view and comfort."[51] And Warren Giles, believing the drama was at an end, announced that the decision "relieves the National League of the necessity of dealing with one of the most serious problems confronting us in many years because I could foresee no possibility of continuing to play in the Coliseum beyond 1959."[52]

The prospect of an extended stay for the Dodgers in the Coliseum was raised even during the celebration over the state supreme court's decision. In a review of the ruling, Frank Finch asked of the contracts' opponents, "How much convincing do those other guys need?"[53] The answer was that they intended to appeal to the United States Supreme Court, as Julius Reuben had predicted even before Judge Praeger issued his ruling. Phill Silver said he would file a petition for review with the United Supreme States Court, and pointed out that the state supreme court opinion was not final for another sixty days and that he could delay filing for another thirty days beyond that.[54] The 1959 baseball season might open before the petition for review was even submitted. Silver raised a number of points on which, he contended, the California court had erred, while City Attorney Arneburgh praised the opinion for its definitive disposing of the issues. Even in defeat, Silver made the most pertinent observation: "This statement by O'Malley that he is going to start building in 30 days is ridiculous. Any action taken before the U.S. Supreme Court acts will be entirely at his peril."[55] In other words, the city and the Dodgers were legally free to commit their resources to the construction of the stadium, but if the Court reversed the appeal or even remanded the case for further consideration, the investment of many millions of dollars might be irretrievably lost.

According to the normal appeal process, the Supreme Court would not announce if it would even hear the appeal until October 1959. The chances were good that the case would not be considered at the federal level, and that, if it were, the judgment sustaining the contract would be upheld. The real issue was time. Were the city and the Dodgers inclined to gamble in order to have the stadium ready for opening day in 1960? After the Dodgers' contract was sustained by the California court, the city exercised further restraint, which placed additional burdens on Walter O'Malley. Condemnation proceedings had been under way against the owners of twelve plots of land in Chavez Ravine totaling 1.36 acres, but once this land had been transferred to the Dodgers, the city dropped its case, leaving O'Malley to reach an agreement with the owners in a series of private transactions.[56]

Once again the difficulty of securing land for stadiums through the private market became evident. Three virtually adjacent plots of identical size (3,924 square feet) were sold for drastically different amounts. One cost $9,312.56, another $18,024.31, and the third $27,637.28.[57] Ironically, these plots, along with four others, constituted the part of Chavez Ravine that the Dodgers were required to set aside as a public playground for local residents. One fortunate owner of a single plot measuring 6,228 square feet found that his land fell at an entrance to the proposed parking lot for Dodger Stadium, which raised the price to more than $150,000. Privately purchasing the remaining land at Chavez Ravine cost the Dodgers just under $500,000, an unanticipated addition to the $15 million needed for construction of the stadium.

On a more nostalgic note, the Dodgers received the good news in January that Zack Wheat had finally been voted into baseball's Hall of Fame.[58] From 1909 to 1926, Wheat appeared in more games than anyone else in the team's history. He also compiled more hits—2,804—

than any other Dodger, as well as more doubles and triples. His life-time average was .317, and in 1918 he led the National League in hit-ting, with a .335 average.

For a team coming off a disappointing season, the Dodgers made re-markably few changes in personnel. They opened the 1959 season still dominated by the veterans of the Brooklyn years, with Snider and Hodges expected to lead the offense and Drysdale, Podres, and Koufax the keys to the pitching staff. The only significant trade had been made during the winter, and it was not expected to produce great dividends. On December 4, the Dodgers traded Gino Cimoli to the St. Louis Car-dinals for outfielder Wally Moon and a relief pitcher, Phil Paine.

At the time, the swap of outfielders seemed even. The key to the agreement, in fact, was the Cardinals' willingness to include Paine. As the *Times* reported, "Not until Devine [the Cardinals' general manager] sweetened the kitty with Paine . . . would the Dodger general man-ager okay the deal."[59] Cimoli and Moon were both in their late twen-ties, and both had had successful years before falling off in 1958. In 1957, Cimoli had appeared in 142 games, batting .293, with 10 home runs and 57 runs batted in. During the first year in Los Angeles, Cimoli was used less frequently, and his performance suffered. In 109 games he hit .246, with 9 homers and 27 r.b.i.'s. His inability to provide more power led Manager Alston to platoon him frequently. Moon had al-most identical figures for the 1957 season, with the exception that he hit 24 home runs and drove in 73 runs. In 1958 Moon also fell out of favor with his manager, and his numbers declined significantly. He appeared in 108 games and had a .238 average, 7 homers, and 38 r.b.i.'s.

The Dodgers' lineup Opening Day included two rookies, Ron Fairly, who started in right field in place of the injured Carl Furillo, and Jim Baxes, who replaced Dick Gray at third base. Snider remained in center field, Hodges was at first base, Charlie Neal continued at second, Don Zimmer played short, and John Roseboro handled the catching. Vet-eran Jim Gilliam was soon to replace Baxes at third, which meant the club had obviously stopped well short of the youth movement that supposedly would purge its lingering identity as the Brooklyn Dodg-ers. In preseason speculation, Frank Finch, an avid Dodger booster even before the franchise shift, picked the club to finish fourth.[60] He pointed with some encouragement to Snider's recovery from a knee injury and the reduced distance to the right-field fence; extolled Drysdale as "the best pitcher in the league" after Warren Spahn of the Braves; and ob-served that "Wally Moon is going to help the club considerably and there is an esprit de corps existing that was painfully lacking last season when the club forded the Gowanus and went west."[61] He concluded, however, that "there still are holes in the lineup."

The new dimensions in the Coliseum were significantly more accommodating to left-handed hitters.[62] The left-field line and power alley remained intact, along with the notorious screen, but from center field to the right-field line the distances were considerably shortened. Center field was reduced a bare 5 feet to 420, but right-center was brought in from 423 to 375 and the right-field foul pole, which had been 390 feet from the plate, was brought in to 333.

The changes in the lineup and the configuration of the Coliseum persuaded few that the Dodgers would seriously challenge for the National League pennant in 1959. The *Times* ran a United Press International story that described the "man to man betting odds on Broadway."[63] The Braves were favored at 6–5, while the Giants and Pirates were 4–1. St. Louis was 15–1, and Cincinnati faced 20–1 odds. The Dodgers, at 25–1, ran ahead of only the Cubs and Phillies, both counted out of the hunt at 100–1.

Perhaps the most historic change for the team as the 1959 season began was the absence of Pee Wee Reese. Reese had begun his career during the MacPhail era, contributing to the pennant of 1941. After wartime service in the Navy he played a vital role on Branch Rickey's team, not only because of his skill on the field but also because of his leadership in supporting Jackie Robinson's integration of the sport. A Southerner, Reese defused some of the racial tension that plagued the Dodgers during Robinson's early career.[64] In 1958, Reese had appeared in only 59 games, compiling an average of .224, as Alston experimented with Don Zimmer and others at shortstop. After sixteen seasons it became clear that Reese's career was at an end. He had appeared in more contests as a Dodger—2,166—than any player but Hall of Famer Zack Wheat. In 1984, his contribution was recognized by the Veterans Committee in Cooperstown, and Reese joined his teammates Robinson, Campanella, and Snider in the Hall of Fame.

The Dodgers' season began in Chicago, where the Cubs defeated Don Drysdale 6–1 in forty-two-degree weather.[65] The game had actually been delayed for a day because of snow, and when bad weather plagued other Midwest openers questions were raised about the sense of having West Coast teams playing on the road in the early spring. In the final game of the series with the Cubs a home run by rookie Don Demeter propelled the Dodgers to victory and they returned to Los Angeles for their home opener on April 14. Before 61,552 fans, a record night-game crowd for the league, they lost to St. Louis.

The local media were already impatient. Paul Zimmerman began his

Times column with the caustic observation, "Let's see now. Last year the explanation of the Dodgers' early season plight included such talk as the uncertainty of the Chavez Ravine situation, the transfer to Los Angeles, etc., etc. [That] now seems to be as good as in the bag and our lads are pretty well housebroken here, but their opening game was hardly an artistic success."[66]

The Dodgers finished April strong, however, and their 11–6 record was good enough to put them atop the National League for the first time since their arrival in Los Angeles. Among their most sparkling performances was a 2–1 win on April 20 in which Drysdale limited the Giants to three hits and struck out eleven. Charlie Neal's ninth-inning homer secured the victory. The next night featured perhaps the greatest hitting performance in the Coliseum's brief history as a major league park. Don Demeter hit three home runs, including one inside the park, leading the Dodgers to an eleven-inning win over the Giants, a nearly four-hour contest ended by Demeter's third homer, which landed in the left-center field seats.

Earlier that day the State Supreme Court unanimously reaffirmed the constitutionality of the Chavez Ravine contract by rejecting the last significant state appeal by the Kirshbaum-Reuben forces. The appeal to the U.S. Supreme Court remained, but for the moment the day was a great success for the Dodgers.

May began with a poignant tribute to Roy Campanella that included an exhibition game with the Yankees before 93,103 fans, the largest crowd ever to witness a baseball game. The receipts were used to benefit Campanella, whose pension status had been clouded, but the Dodgers also profited from the public relations, which could hardly have come at a better time. The stadium controversy was about to take an ugly turn as the last residents of Chavez Ravine were forcibly evicted from the homes they had maintained through years of political controversy and litigation.[67] When county sheriff's deputies arrived with eviction notices, most residents left quietly, but the Arechiga family refused to abandon their house, which sat on property acquired by the city in 1951 for the original purpose of constructing public housing in accord with the Federal Housing Act of 1949. The Arechigas were compensated in the amount of $10,050, and the money placed in an account in their names when they failed to challenge the final judgment. Two years later, once it became apparent that public housing would not be constructed on their site, the family brought suit to overturn the judgment that had taken their homes.

The district court of appeals, ruling on the case in 1957, found that the family no longer held any claim to the land: "The Arechigas make no claim of fraud, collusion, bad faith, or other misconduct on the part of any governmental agency . . . when the judgment in the condemnation case became final, they were divested of all interest in the property regardless of the purpose for which it might later be used."[68]

"From 1953," Norris Poulson added, "the Arechigas lived rent-free on the property. The $10,500 was held available for them, minus a deduction of $11 for unpaid personal property taxes. The city, on second thought, did not care to charge rent for houses which had been declared substandard. The Arechigas had refused to move in 1951 when the resettlement agency was finding dwellings for families whose property had been condemned."[69]

While the controversy over the Dodgers raged through 1958, the family continued to live in Chavez Ravine. Years later, a priest who grew up in that setting recalled, "With all the studies I see about what people need to grow up healthy, what psychologists say, that's what we had."[70] He mentioned warm relations within and among extended families living in peace and security. "I have beautiful memories of our childhood," he concluded. On May 8, 1959, as deputies forcibly removed the Arechigas from their dwelling, Mrs. Avrana Arechiga, the sixty-eight-year old matriarch, threw rocks at them, while her daughter, Mrs. Aurora Vargas, a war widow, was carried kicking and screaming from the premises.[71] Mrs. Victoria Augustain also physically resisted the eviction, while children cried and pets, chickens, and goats added to the chaos. The grim scene was televised by local stations in the city.

A Spanish-speaking deputy tried to reason with the family, who had barricaded themselves in their home, but the exchange turned into a shouting match. A deputy tore out a window through which movers transported furniture. When Mrs. Augustain left the house for a television interview, she was blocked from returning. Shortly after Mrs. Vargas was removed, leaving the houses vacant, bulldozers tore the structures to the ground. The furniture was taken away to storage, but the family literally pitched a tent in a vacant lot nearby, claiming they had nowhere else to go.[72] In the evening, Edward Roybal, the Arechigas' city councilman, arrived to see if the family needed food or other provisions. "The eviction in itself is legal," Roybal commented the next day, "but the manner in which it was carried out certainly was not. Someone stands to answer for violating the individual rights of these people. This is the type of action that occurred during the Spanish Inquisition and Hitler's Germany. But never have I heard of anything like this taking place in this city."[73]

As if the family hadn't enough problems, Mrs. Vargas was charged with battery for resisting the deputies, and she later claimed she served thirty days in jail and suffered a nervous breakdown as a result of the ordeal.[74] "It was natural to have felt sympathy for the Arechiga family," Poulson himself later remarked, "while viewing pictures of the eviction."[75]

Public sympathy vanished several days later when a local paper reported that the Arechigas were not as destitute as they appeared: "The 'homeless' Arechiga family, 11 of them living in a tent to dramatize their eviction from Chavez Ravine, owns at least 9 homes, the *Mirror-News* learned today. The Arechigas . . . rent most of their properties at prices up to $75.00 a month. Although Abrana, 69, and Manuel Arechiga, 72, have been living in the canvas shelter five days, a son-in law, Miguel Augustain, owns a vacant three-bedroom house a short distance away."[76]

After the revelations about the Arechigas' financial assets, Mayor Poulson characterized the entire affair as a television tactic for cracking the hold on local news coverage previously enjoyed by the newspapers: "To break this spell, it would be natural that Television would resort to sensationalism—and I can say half-truths. A half true is more destructive than an outright lie."[77]

After acknowledging the natural sympathy anyone would feel for the Arechigas, Poulson returned to his theme of the pernicious effects of television: "Sympathy today is misplaced. A community must live by laws. The Arechigas have chosen to flout the law. The petition calling for justice seeks to make a mockery of the law. When the Arechigas were carried struggling from their homes, they were fully aware the cameras were trained on them. For the past week they have been all three rings of a three ring circus, especially when they lived in tents. No doubt they enjoyed every indignant minute of it."[78]

Not surprisingly, the newspapers approvingly quoted Poulson's attacks on Roybal and their competitors. The *Times* wrote: "If the government is to see its legal decisions tested in the streets—tested by ham actors on television—tested by the strident voices of City Councilmen seeking to gain the limelight—and not tested in the courts where they should be, we must admit it is a sad day for America."[79] The newspaper went on to declare:

> The television pictures—wonderful action pictures—were the answer to a demagogue's prayer. With such pictures facts would only spoil the effect. The television commentators shared this view. There may never have been a better example of the emptiness of on-the-spot television news coverage. The commentators or reporters—actors was the mayor's felicitous word— were like the cameras; they could see, but they could not think or ask questions.

How many people who look and listen but only read while they run will ever be cured of the opinion that the Arechiga clan was turned out of doors without decency or fair warning? For all of these misled people, blame the television news—actors, who bled insincere tears up and down the picture tube without any effort to find out or present the true facts.[80]

Councilman Roybal attempted to argue that the Arechigas' other holdings changed nothing about the manner of their eviction, but public attention had turned elsewhere. A family uprooted for the sake of a stadium compelled interest, especially when television cameras presented the spectacle of uniformed law officers carrying women from their homes while children screamed in horror. Similarly, when the stories broke about the other property the family owned, few were prepared to follow closely reasoned arguments about the appropriate method of evicting a family. Once the public was informed that the family could care for its elderly and children, the drama ended. The market for more information had evaporated, and the focus of print and electronic media turned to other, more sensational matters.

Three perspectives are offered by men who played a role in the Arechiga cases. Edward Roybal, now a congressman from Los Angeles, remains disturbed about both the treatment of the family and the contract which transferred Chavez Ravine to the Dodgers. He maintains that the episode has left a residue of bad feeling among his constituents in the Hispanic community of Los Angeles.[81] Roger Arneburgh, now in private law practice, insists that there was no recourse to evicting the families. He concurs with Poulson that television coverage sensationalized the removal of people who had no legal right to be living in Chavez Ravine.[82] Kenneth Hahn, who continues to serve on the County Board of Supervisors, allows that eviction was legal, but that from a politician's perspective the forced removal of elderly residents from their homes is a measure to be avoided at virtually all costs.[83] He adduces his ability to persuade several hundred families to vacate their homes for the construction of a hospital as evidence that persuasion can obviate the need for forced eviction.

The Dodgers struggled through the remainder of May, losing five in a row in the middle of the month and dropping to fourth place. They rebounded in June, although they briefly fell to fifth place before a seven-game winning streak at the end of the month brought them within a game of first place. There were further transitions. On June 15, Carl Erskine announced his retirement. "Oisk," as he had been known in Brooklyn, had struggled with arm trouble for several years, after winning twenty in 1953 and pitching two no-hitters.[84] Roger Craig was brought up from Spokane as a replacement, and the Dodgers also looked to their farm system to bolster the shortstop position, where Don Zimmer and Bob Lillis had provided spotty offense as replacements for Pee Wee Reese. The team called up a twenty-nine-year-old rookie, Maury Wills, who had proved capable with his glove in Triple A though his offensive contribution was limited to stealing bases.

June ended amid some controversy. On the night of June 30, Sam Jones of the San Francisco Giants held the Dodgers hitless for seven and two-thirds innings. Jim Gilliam then hit a slow ground ball to short, which Andre Rodgers juggled. The official scorer ruled the ball a base hit, to the consternation of the Giants, and a crowd of over fifty-nine thousand viewed with mixed emotions as the Dodgers escaped an embarrassing contribution to history.

The first half of the season ended with the Dodgers in third place, half a game behind the Giants, with the Braves also in close contention. Don Drysdale opened the All-Star Game for the National League, pitched three perfect innings, and was voted the game's Most Valuable Player. A second All-Star game was scheduled later in the season to fatten the wallets of the owners, who wanted to exploit the seating capacity of the Coliseum. Good sense was restored in 1963, when the Dodgers left the cavernous stadium and the leagues reverted to a single summer classic. Roger Craig's four-hit win over the Pirates on July 29 briefly put the Dodgers back in first place. Two days later, they moved within half a game of first, as Drysdale won his sixth straight, striking out fourteen. Drysdale started the second All-Star game, held on August 3 in the Coliseum, where 55,105 watched him give up three runs as the American League won 5–3. The turnout for the All-Star game might have been disappointing, but the local fans compensated by showing up in force for a more meaningful contest—ninety thousand saw the Dodgers defeat the Braves, 4–2, on August 8.

That month the team fell back to four and a half games out on August 24, when at last Sandy Koufax began to display the greatness that had been expected of him. He hurled the team into second place by striking out thirteen Phillies, the first of three record-setting outings. On August 28, Koufax tied a modern record by fanning eighteen Giants, and Wally Moon hit a three-run homer in the ninth to defeat San Francisco 5–2. More than eighty-two thousand were in attendance, as the fierce race for the flag drove the city into its first major league pennant fever. Koufax's next outing was on September 6 against the Cubs. He struck out ten, establishing a new three-game strikeout record, eclipsing the old mark first set by Walter Johnson and later tied by Bob Feller. Subsequent performances by pitchers like Steve Carlton and Nolan Ryan have accustomed the modern fan to spectacular strikeout marks, but it was Koufax who pioneered this trend. He also helped mold the distinct character of the next great Dodger teams, which, rather than relying on the powerful bats that had dominated the National League in the 1950s, emphasized pitching.

The Dodgers entered the home stretch with strong pitching but a makeshift lineup. Snider, Hodges, and Furillo continued to contribute, but their injuries forced the team to rely on other players. First, Don Demeter came through; when he cooled off after his hot start, Norm Larker, Ron Fairly, and John Roseboro picked up the slack. Maury Wills played well in the field, but as of September 7th he was batting an anemic .207. The most pleasant surprise was Wally Moon, who had indeed added power to the Dodger lineup, although in a somewhat curious manner. As of September 1, Moon had hit eight home runs in the Coliseum, six of them over the left-field screen—"Moonshots," as their looping trajectories came to be called in those early days of space exploration. Moon meticulously explained his technique: "If you want to slice the ball, it's simply a matter of bringing the hands closer to the body and slightly delaying the swing. You keep the end of the bat cocked for a split second after your hands have begun to move, and at the last possible moment you flip the end of the bat at the ball. It's actually a deliberate slice."[85]

Moon was not the most dangerous power batter in the lineup that season. In 107 fewer at bats, Duke Snider hit 4 more home runs than Moon's 19 and Gil Hodges led the club with 25 in 130 fewer trips to the plate than Moon. The important contribution of the Dodgers' new left-fielder was that his success to the opposite field gave pitchers something new to think about when facing the Dodgers lineup. Moon also gave sportswriters and fans something new to discuss. He helped establish the new identity that distinguished the team from its Brooklyn ancestors. Hodges and Snider were familiar stars of a club still associated with Ebbets Field, but Wally Moon, by way of the St. Louis Cardinals, was the first star of the Los Angeles Dodgers.

On September 17, the Dodgers concluded their final home stand tied with the Braves for second place, two games behind the Giants, whom they would face in a three-game series scheduled for Seals Stadium in San Francisco. September 17 was also the day that ground was broken for Dodger Stadium in Chavez Ravine.[86] Work commenced despite the appeal pending with the Supreme Court, which might announce the fate of the ballpark in another month. Various local dignitaries joined Walter O'Malley in turning the first shovel of earth before bulldozers began the serious labor.

The Dodgers then flew north for their most important series since leaving Brooklyn. With only five games left in the season they needed a sweep of the three games to assume first place in the league. In that unlikely quest, the Dodgers were relying more and more on Maury Wills, who not only had begun to hit for average, but was flashing the skills that would restore base stealing as a major dimension of the game, although he would finish the season with only seven stolen bases.

Nature aggravated the tension by raining out the series opener. Day and evening games were scheduled for September 19, as Horace Stoneham declined to let capacity crowds fold into a doubleheader. In the afternoon game, Wills went three for four, scoring two runs, while Roger Craig gave up only one run—in the ninth inning—and pitched the Dodgers to a 4–1 victory. In the nightcap, the Dodgers scored five times in the seventh to take a 5–3 win. In the decisive frame, Wills singled but the key hit was a double by Charlie Neal. When news of Neal's hit was announced in the Coliseum during the Rams' game with the Philadelphia Eagles, play had to be stopped until the noise subsided enough for the quarterback's plays to be heard. In the final game, Johnny Podres faced twenty-game winner Sam Jones, who had narrowly missed no-hitting the Dodgers earlier in the season. Duke Snider hit a solo home run in the second, and Jones walked in Maury Wills two innings later. The Dodgers scored twice more in the seventh. The Giants halved the lead in the eighth, and Larry Sherry and Sandy Koufax appeared in relief to halt the threat. Four runs in the ninth iced the victory for the Dodgers and propelled them into first place for the first time since the end of July.

The sweep provided vital momentum to the team and gave a foretaste of the Dodger teams of the 1960s, which combined great pitching with Wills's offense. Even playing in a minor league park, the powerful Giants lineup produced only six runs in three games. A telling moment occurred in the first inning of the second game, when Drysdale walked the bases loaded, but then struck out in order Orlando Cepeda, Willie Kirkland, and Daryl Spencer. Wills went seven for thirteen for a .538 average, and he scored in all but two of the innings in which the Dodgers tallied runs. He was still batting near the bottom of the order, but soon he would become the consummate leadoff man. For the Dodgers' old guard, the series was less successful. Gil Hodges was hitless in fourteen at bats, while Snider and Furillo had only six plate appearances. Snider's home run demonstrated that they remained important figures; but injuries were slowing them down and other players were filling the void.

After their triumph in San Francisco, the Dodgers flew to St. Louis, where they dropped the first of a two-game series and yielded first place to Milwaukee. A shutout by Roger Craig, combined with Pittsburgh's win over the Braves, put the Dodgers back in a tie for the lead, while the Giants fell two games behind with only three left to play. The Dodgers concluded the season in Chicago, while the Braves finished at home against Philadelphia. In the first game of the final series, Gil Hodges shook off an arm injury and beat the Cubs with an eleventh-inning homer. The Braves, meanwhile, were upset by the last-place Phillies and slipped to one game back.

On Saturday, the Braves regained a tie with the Dodgers, as Warren Spahn pitched a 3–2 win over the Phillies while the Dodgers were blown out by the Cubs, 12–2. On the last day of the season, Roger Craig took the mound and hurled the Dodgers to a 7–1 win. Of their final eight games, Craig had won three, yielding a total of two runs and becoming the team's most effective pitcher as Drysdale and Koufax both struggled. The Braves gained a tie for the pennant when the Phillies showed why they had secured last place. Four errors, a balk, a wild pitch, and a passed ball led to five unearned runs as the Braves won 5–2 to force the third playoff in National League history, all of them involving the Dodgers, who had lost to the Cardinals in 1946 and to the Giants in the famous playoff of 1951. This particular method of deciding pennants must have caused some Dodger fans to favor a coin toss instead.

The three-game playoff began in Milwaukee on Monday, September 28. Larry Sherry a twenty-four-year-old native of Los Angeles who had compiled a 7–2 record for the Dodgers since being called up from Spokane in July, relieved Danny McDevitt in the second inning and pitched seven and two-thirds innings of shutout ball, nailing down a 3–2 win. John Roseboro homered in the sixth inning. No less than Roger Craig, Sherry was emerging as a star. He lost his first two starts because of tough luck, then reeled off seven wins in a row while compiling an e.r.a. of 2.20. He was used exclusively in the bull pen in the last few weeks of the campaign, but his experience as a starter was evident in the long appearance against the Braves. Sherry's success diminished the significance of Clem Labine, the Dodgers' stalwart relief pitcher during their final years in Brooklyn. Here was more evidence that Walt Alston was turning to fresher faces to play vital roles in the pennant chase.

On Tuesday, the playoff shifted to Los Angeles for the final two games. Don Drysdale faced Lew Burdette in the second game, now a must win for the Braves. Although the Dodgers had beaten Burdette eleven times in thirteen games in County Stadium, "Fidgety Lew" had triumphed in his only two starts in the Coliseum in 1959 and had won nineteen more during the campaign. The Braves broke on top with two runs in the first inning. The Dodgers retaliated once in the bottom of the inning, but the Braves rebounded with another run in the second. The Dodgers added a run in the fourth inning, but in the fifth the Braves matched them again. When Milwaukee scored again in the eighth, they led 5–2, with Burdette in control of the game.

Entering the bottom of the ninth the Dodgers trailed by three runs and faced a third and final game on Wednesday. But the old Brooklyn stars took over. After Wally Moon opened the ninth with a single, Duke Snider's base hit brought the tying run to the plate in the person of Gil Hodges. Hodges stroked a single to left, and Pee Wee Reese, the Dodgers' third base coach, held Moon at third. The bases were loaded with no one out. Don McMahon replaced Burdette, and gave up a single off the screen to Norm Larker, scoring both Moon and Bob Lillis, who was running for Snider. Warren Spahn came in to face the left-hand-hitting catcher John Roseboro, and Walter Alston countered by sending Carl Furillo to the plate.

Furillo had appeared in only fifty games that year, and as a pinch hitter in half of them. On the second pitch from Spahn, Furillo drove the ball deep to right field, where Henry Aaron made a one-hand catch. Gil Hodges tagged up to score the tying run. Later in the inning, with the winning run on third, Jim Gilliam also hit the ball deep to right, and Aaron again made a fine catch to send the game into extra innings.

Sherry was spent from his outing the day before, so Alston called on Stan Williams to stifle the Braves. Although he walked three batters, Williams held the Braves hitless in the tenth, eleventh, and twelfth innings. To open the twelfth, Wally Moon popped out and Williams flied to center and the marathon game was approaching four hours with no end in sight. Hodges again came through by drawing a walk to keep the inning alive. Then Joe Pignatano, the Brooklyn-born reserve catcher, singled, advancing Hodges to second. Again Carl Furillo, the only veteran of the Dodgers' playoffs in 1946 and 1951, came to the plate.

In those previous series Furillo had totaled one hit in twenty-three at bats. Facing relief pitcher Bob Rush, Furillo hit a grounder up the middle barely gloved by shortstop Felix Mantilla, who had started the game at second base. With Furillo running on bad legs, Mantilla thought he could end the inning by throwing to first base. His throw bounced in the dirt, hopped past first baseman Frank Torre, striking Greg Mulleavy, the Dodgers' first base coach, and caroming into the stands. Waved home by Reese, Hodges crossed the plate to give the Dodgers their first pennant in Los Angeles.

Walter Alston used twenty players to secure that final victory. His exhausted team had only a day to celebrate and travel to Chicago for their first World Series game against the White Sox, who had won ninety-four games that year, taking their first pennant since the Black Sox scandal of 1919. They were led by an outstanding pitching staff, and had good speed and defense, especially in their double-play combination of Nellie Fox and Luis Aparicio. In the first game of the Series in Comiskey Park, the Sox pounded Dodger pitching for an 11–0 win. Roger Craig was lifted during a seven-run third inning. In the second game, after the White Sox scored twice in the first inning, the Dodgers rallied by playing long ball. Charlie Neal hit a solo homer in the fifth and another with a man aboard in the seventh to give the Dodgers a 3–2 edge. Chuck Essegian, like Larry Sherry an alumnus of Fairfax High in Los Angeles, pinch-hit a home run, also in the seventh, to provide insurance as Sherry, yielding one run in the eighth, saved the game for Johnny Podres.

After a travel day, the third game was played in the Coliseum before 92,294 fans. Carl Furillo was again the hero as his bases-loaded pinch-hit single in the seventh broke up a scoreless pitchers' duel between Drysdale and Dick Donovan. Larry Sherry pitched the final two innings as the Dodgers won 3–1. In the fourth game Roger Craig rebounded, carrying a 4–0 lead into the seventh inning, when the Sox tied the game, aided by catcher Sherman Lollar's three-run homer. Sherry blanked the Sox in the eighth, and Gil Hodges provided another bittersweet moment for Brooklyn fans with a homer that gave the Dodgers and Sherry a 5–4 win and a three-to-one-lead in the Series. A record 92,706 turned out to urge the Dodgers to a world championship. Sandy Koufax, dueling Bob Shaw, pitched brilliantly, holding Chicago to five hits and one walk while striking out six in seven innings. But he was undone in the fourth when the Sox loaded the bases with one out and scored the only run of the game on a double play.

On October 8 the Series returned to Comiskey Park for the sixth game. Early Wynn opposed Podres, who was trying to win his and the Dodgers' second world championship. Snider staked him to a 2–0 lead with a homer in the third, in the fourth the Dodgers scored six more times and the game seemed out of reach. The Sox rallied in the bottom of the fourth with a three-run clout by Ted Kluszewski, his third home run of the Series. It chased Podres and Alston again called on Sherry, who pitched the final five and two–thirds innings. Essegian pinch-hit another home run in the ninth to set a Series record, and Sherry retired the White Sox in the bottom of the ninth to gain his second Series win to go with his two saves. In only their second year, the Los Angeles Dodgers won their first World Series championship.

The 1959 Dodgers were not among the great teams in baseball history, but they bridged two great teams, the Brooklyn Dodgers of the 1950s and the Los Angeles Dodgers of the 1960s. The significance of the 1959 championship was that it legitimated the franchise shift and clarified the controversy surrounding a new stadium in Chavez Ravine.

"Who will have the effrontery to tell us now—the several million of us—that the movement to supply the Dodgers with a decent playing yard was against the public interest?" the *Los Angeles Times* exulted in an editorial.[86] "Their triumph is that they have created one of those centers of attachment that the metropolitan area of Los Angeles has needed so desperately. The team has made the people for a couple of hundred miles around aware that they have a common interest. A major league baseball club does not a city make, but in our agglomeration of Southern California communities any joint enterprise which excites a wide interest serves as a sort of civic glue."[87]

The White Sox and many observers of the Series were highly critical of the Coliseum as a site for the ultimate showcase for baseball, but gate receipts aside, no one had ever claimed that the Coliseum was the reason for transferring the Dodgers from Brooklyn. It was simply a stopgap until the stadium O'Malley had wanted for years could be constructed. If all had gone as planned, 1959 would have been the final year for the Dodgers in the bizarre field, and if the team had not been so resourceful, the public spotlight would not have shone in October on the distorted Olympic stadium.

The 1959 season remains a charming oddity in the history of the Dodgers, but another October victory had a more permanent impact. On October 19, the United States Supreme Court dismissed the appeals of Louis Kirschbaum and Julius Reuben, ending the judicial challenges to the Chavez Ravine contract. By exercising its discretion to reject the appeal, the Court affirmed the rulings of the previous January by the California State Supreme Court. The final obstacles were removed and the construction begun in September could continue unimpeded.

In contrast to the joy and pride in Los Angeles, the mood in Brooklyn after the Dodgers' win was subdued. Four groundskeepers and watchmen viewed the final game on television in a maintenance room in Ebbets Field and reported that they received only a few phone calls after the final out. Perhaps the feelings of the borough were summarized by one fan who said, "We were rooting for Furillo and Hodges and Snider, those other guys—Wills, Demeter and all—we didn't root for them. . . ."[88]

9

Was the Move Justified?

On April 10, 1962, more than a decade after Walter O'Malley initiated his pursuit of a new ballpark, Dodger Stadium opened in Los Angeles. In that setting, 3-million attendance seasons have become routine; in season ticket sales alone, the Dodgers draw two million per year, exceeding their total yearly sales in Ebbets Field. During its tenure in Los Angeles, the team has won eight National League pennants, compared to nine during their almost seven decades in Brooklyn; and four world championship flags have flown in Los Angeles compared to the lone banner in Ebbets Field. By all measures, the Dodgers have been a spectacular success in Los Angeles, but a question continues to plague the franchise: was the move justified?

For Roger Kahn, Peter Golenbock, and others who followed the team in Brooklyn during its final years, the move represents the unforgivable abandonment of a community that had supported the team for sixty-seven years. Kahn maintains that a stronger commissioner would have blocked the move, and Bowie Kuhn, the former baseball commissioner, doubts he would have allowed the transfer if it had been attempted during his tenure.[1] For Dodgers' supporters in Los Angeles, however, the move helped make baseball a truly national sport, and Los Angeles can justly claim that it has surpassed any city in the consistent backing it has given its team over thirty years.

The dispute about the propriety of the Dodgers' move persists not only because of the intensely conflicting emotions of the fans in Brooklyn and Los Angeles, but also because several distinct aspects of franchise relocation have become muddled. By treating these issues separately, each from its own perspective, we can gain a clearer understanding not only of the Dodgers' case but also of subsequent moves.

The first perspective is that of *business*, in which the Dodgers must be seen as an enterprise in an economic market. Almost every franchise has abandoned stadiums constructed at the time Ebbets Field was built, and the few that haven't—the Red Sox, White Sox, Tigers, and Cubs—are facing strong incentives to move to more modern facilities. The Dodgers could not have remained in Ebbets Field indefinitely, even if Walter O'Malley had wanted to. The only alternative for the Dodgers, if they wanted to stay in New York, was to accept the kind of arrangements the Mets have secured: tenancy in a public stadium in Queens. Robert Moses's objections to a privately financed stadium and the futility of the Brooklyn Sports Center Authority precluded the use of the Atlantic-Flatbush site; no other location in Brooklyn was feasible. Thus at some point in the 1960s, the Brooklyn Dodgers would have ceased to exist, even though they might have remained in New York. The most optimistic speculation about their future in New York pales beside the success they have enjoyed in Los Angeles. From the perspective of a firm seeking to maximize its return on an investment, the Dodgers made the correct decision when they moved to Los Angeles.

The players' two-day strike in 1985 made it possible to examine the finances of major league teams because the owners released statements to back their claim that under the combined effects of arbitration and free agency most clubs were losing sums so substantial that they threatened the entire financial structure of the game. In response, the Major League Baseball Players Association hired economist Roger Noll of Stanford University to evaluate the statements of the twenty-six major league teams.[2] Noll found that the statements tended to distort the data because almost all the teams are part of large corporations, such as broadcasting or shipping firms, which can obscure the actual dollars involved in baseball operations.

The Dodgers, the only major league team whose ownership is involved in no business except baseball, declared that their net revenues were over $6 million. But Noll believes that "these profits, richly deserved as they are for the best managed team in sports, vastly understate the value of the Dodgers to the owners."[3] Noll concludes that the Dodgers "are probably the most successful sports franchise that has ever been fielded." And in a trenchant summary that evaluates each of the major league teams in economic terms, the Dodgers are described succinctly as "baseball's answer to the Denver mint."[4]

The Dodgers would have retained substantially the same organization even if they had remained in New York, but undoubtedly a critical component of their prosperity is their stadium, which is not only the best in the game but also the only modern structure owned free and clear by the team. When Walter O'Malley died on August 9, 1979, ownership of the team was bequeathed to his son, Peter, and his daughter, Terry. Peter O'Malley has continued his father's style of management, which treats the Dodgers as a business that stands or falls on the skill and attention the executives and employees bring to their jobs. The O'Malleys have treated the baseball team as an enterprise that must be promoted through sensible investment and constrained through fiscal prudence. An article in *Forbes* in 1982 considered the ingredients, other than fielding quality teams, which contributed to the Dodgers' commercial success. "The Dodgers are the best-run franchise in all of baseball because of a lot of other things they do right. Compare Dodger Stadium, for example, to another entertainment success of the West Coast—Disneyland. Dodger Stadium is squeaky clean, beautifully landscaped and rests in a striking setting. As at Disneyland, Dodger Stadium attendants—even in the parking lot—are civil. The bathrooms are clean and safe."[5]

"Do people go to Dodger Stadium to behold flowers?" asked Roger Kahn, commenting on their presence at Dodger Stadium. "Not primarily, but people have stayed away from other ball parks because of ambient filth."[6] The organization has been likened to IBM, but a more apt analogy is a small family business that thrives because of its attention to detail and its awareness that shifts in the market, if not addressed, can become fatal. Since 1939, under MacPhail, Rickey, and the O'Malleys, the Dodger organization has kept pace with social and economic changes that affect baseball, and this has been a vital element in the club's financial success.

Two changes that confront the Dodgers presently are the escalation of players' salaries and the related matter of dilettantish ownership committed to a win-at-any-cost strategy. Peter O'Malley is troubled by both developments, but he focuses on "the system" of free agency, which he believes is responsible, in combination with arbitration, for perilously increasing labor costs.[7] The desire of owners to limit free agency and arbitration has led to confrontations with the players' union that resulted in the extended strike of 1981 and the two-day strike of 1985. O'Malley expresses confidence that his family-owned franchise can prevail against corporate competitors, but he is clearly alarmed by what he considers threats to the long-term financial health of the game resulting from wildly escalated salaries. Since he must generate the revenues to pay these salaries, O'Malley is understandably concerned, but he underrates another more fundamental point. Salaries are ultimately determined by owners in response to market demands. If their decisions are irresponsible the fault hardly lies with the players, their union, or even with an amorphous "system." Inescapably, one must conclude that the owners' problems are self-inflicted. The Dodgers have been comparatively restrained in their personnel decisions, though they are inevitably affected by arbitrators' decisions that reflect, in part, the salary choices of other teams.

The Dodgers' model of family ownership has become increasingly rare in baseball. In recent years, the Stonehams, Griffiths, Carpenters, Galbreaths, and Wrigleys have sold their teams to new owners without long-standing commitments to baseball. Certainly these sales have been influenced by the difficulty family owners have keeping pace with the increased costs of operating a baseball franchise. The argument can be made that this change in ownership has benefited the game by infusing more money into it, and that as long as those who run ball clubs exercise sound judgment the new style of ownership serves a useful purpose. Bowie Kuhn points to the Chicago Cubs as a franchise that has prospered under corporate ownership, and Peter O'Malley notes that corporate ownership may reduce the extremely idiosyncratic behavior sometimes exhibited by individual owners.[8]

The key point is that increased revenues have not made baseball financially more secure. The greater sums are proper targets for the players and their agents, and as the sources of even higher salaries. Teams in smaller markets like Pittsburgh and Seattle eventually will face even greater strains because of the higher level at which equilibrium is pursued. Financial security will be reached only when costs and revenues achieve a reasonable balance. As long as some teams are run as hobbies or diversions, prudent business decisions will be virtually impossible, and every club will eventually feel the pressure. The Dodgers have been run as a serious business since the time of Larry MacPhail. Their subsequent prosperity is no coincidence.

The owners, having at first been intoxicated by free agency, are now abstaining. For the past two years, clubs have collectively declined to bid on free agents and have lowered their rosters to twenty-four. Even if these measures withstand changes of illegal collusion, they patently deny the players' lawful option of a free agent market. The owners have yet to prove that they can operate their firms in a prudent manner while respecting the players' legal rights. The lesson professional football has taught is that owners have more to fear from one another than from the players or their unions, and eventually an owner who believes his team is one player away from a championship will likely embark on another spending binge. When the basic agreement between owners and players expires on December 31, 1989, the owners

may once again assume their curious position of insisting that the players stop them before they spend again.

The second perspective from which to evaluate the move to Los Angeles is the *romantic* attachment of the fans. From this perspective the
move can never be justified to the Brooklyn Dodgers' fans. Revenues,
tax advantages, mythical mineral rights, demographic shifts, and other
market forces are utterly irrelevant. What counts, as Roger Kahn has
demonstrated so well, are the bonds developed over generations. The
memories shared among friends and families who exchange stories of
"Uncle Robbie," Zack Wheat, Casey Stengel, Babe Herman, Leo Durocher, Gil Hodges, Carl Erskine, and Jackie Robinson don't compute
on a balance sheet. These memories not only remain a vital part of
Brooklyn itself; they extend to communities where former Brooklyn
residents themselves have moved, and have become a part of baseball
folklore.

Describing the frustrations of the Dodgers trying to wrest a world
championship from the Yankees in the 1950s, Roger Kahn has written,
"You may glory in a team triumphant, but you fall in love with a team
in defeat. Losing after great striving is the story of man, who was born
to sorrow, whose sweetest songs tell of saddest thoughts and who, if
he is a hero, does nothing in life as becomingly as leaving it. . . . A
whole country was stirred by the high deeds and thwarted longings of
The Duke, Preacher, Pee Wee, Skoonj and the rest. The team was awe-
somely good and yet defeated. Their skills lifted everyman's spirit and
their defeat joined them with everyman's existence, a national team,
with a country in thrall, irresistible and unable to beat the Yankees."[9]

The reaction in Brooklyn to the Dodgers' move is understandably
bitter. Among the more moderate reflections recorded by Peter Golenbock from the abandoned fans of Brooklyn is this one: "The day it was
announced, if you were in Behan's Bar and Grill, you'd have thought
it was a wake. This was like seceding from the Union."[10] "After all
those years of heartbreak," said another fan, "to have it end like that—
I never forgave the National League. I have never gone to a National
League game since."[11] And another: "When the Dodgers left, it was
not only a loss of a team, it was the disruption of a social pattern.
There was no more sense of waiting up for the *Daily News*. The life
went out of the street corners. What were you going to stand there
and talk about?"[12]

The most vitriolic remarks focus, of course, on Walter O'Malley, whose
unbridled greed, it is asserted, was the principal cause of the Dodgers'
move. Golenbock summarizes O'Malley's petitions to Robert Moses and

other city officials in New York as, "Give me a bigger and newer stadium. . . . Give me. Not build me. Give me."[13] Elsewhere, Golenbock quotes former Dodger publicist Irving Rudd:

> Whatever anyone wants to say, New York Mayor Robert Wagner and Parks Commissioner Robert Moses were right. If they had given O'Malley what he wanted, they should have gone to jail. If you remember, the talk was about a Superdome in downtown Brooklyn, where the Long Island Railroad is, with stores, apartments. And who do you think would have owned all of this? O'Malley. He was offered the use of the Flushing Meadow site, a la Shea, and his answer was, "You build a ballpark and *give* it to me."[14]

In this interpretation, passion has overwhelmed the facts. O'Malley was always emphatic about constructing the new stadium with private funds, as indicated by his remarks at the time he announced the Jersey City move:

> To clear up any misconceptions, I would like to make it plain that we are not going into this thing with our hats in our hands.
>
> We are—and have been for some time—ready, willing and able to purchase the land and pay the costs of building a new stadium for the Dodgers. We have $6,000,000 available for this purpose if an adequate site can be made available.
>
> In fact, I feel rather proud to be associated with this evidence that the days of individual enterprise have not ended. This effort by the Brooklyn club is the first by any baseball club in more than fifteen years to build a new plant. . . .
>
> I believe the Dodgers should own their own ball park.[15]

O'Malley's motives for wanting a private stadium were clear in his first public remarks about a new stadium for Brooklyn. In December 1953, announcing his plans for a replacement for Ebbets Field, he said, "The new park will not be a municipal stadium. That would mean a political landlord, which isn't desirable."[16] For O'Malley, saving money in the construction of a new stadium mattered less than gaining control over the new park. The record in these pages establishes persistent efforts by O'Malley to build his own stadium in New York. The Dodger owner may be a convenient target for the emotions of Brooklyn fans, but the real causes of the Dodgers' move were far more complex than those put forward by the O'Malley Devil Theory.

For all the pathos that attends the recollections about the Dodgers in Brooklyn, the borough does not have a monopoly on the romantic perspective. In spite of popular impressions of Southern California, the Dodgers have established roots and traditions in Los Angeles similar to those recorded so poetically by Kahn. For thirty years, the Dodgers have continued to play a special brand of baseball unaltered by their new surroundings and by players not even born when the team moved West. The essence of the Dodgers' particular style of play may be that, whereas Yankee fans always expect their team to win and Red Sox fans fear that ultimately their team will lose, Dodger fans know that no matter how their club is doing at the moment, it is capable of reversing fortune to pull out either victory or defeat.

This pattern that so characterized the team's final years in Brooklyn continued in the Coliseum and still is in evidence in Dodger Stadium, enthralling millions of fans. The roller coaster began immediately in Los Angeles as the Dodgers rebounded from nearly a last place finish in 1958 to a world championship in 1959.

In 1960, the team slipped to fourth place as it continued to assemble the pieces of the one great Dodgers team that has appeared in Los Angeles. Tommy Davis and Frank Howard joined Wally Moon in the outfield, while Norm Larker took over for Gil Hodges at first base. Ron Fairly served a year in the Army. And the pitching staff posted unremarkable numbers. Maury Wills emerged as a star of the 1960 team, batting .295 and stealing fifty bases to lead the league. Howard's twenty-three home runs won him the league's Rookie of the Year award. Davis's .276 average gave evidence of skills that would lead to two batting titles in the 1960s.

Another link with Brooklyn was cut in 1960 when Carl Furillo, the hero of the playoffs against the Braves, was released to make room for Howard, the rookie sensation. Furillo left bitterly, contending he was being let go because of injuries and that the Dodgers were trying to renege on his salary.[17] Those claims were eventually sustained, and he was awarded his pay, but the matter compounded the awkward, indecorous ways in which the Dodgers tended to sever their Brooklyn associations.

In 1961, the club rebounded to a second-place finish, four games behind the pennant-winning Reds. Wills again led the league with thirty-five stolen bases, while hitting .282. Walt Alston platooned the few remaining Brooklyn stars with the many new players competing for permanent spots in the lineup. Ron Fairly returned from the Army to share outfield duty with Snider, Tommy Davis, Wally Moon, Frank Howard, and a new "phenom," Willie Davis. Sandy Koufax broke through to win a career-high eighteen games, a total matched by Johnny Podres. Don Drysdale finished at 13–10, while Stan Williams was 15–12. Larry Sherry compiled a 4–4 record with a 3.90 e.r.a, but Ron Perranoski emerged from the bullpen in his rookie year to post a 7–5 record with an e.r.a of 2.65.

The inexorable transition continued as Gil Hodges played his final games for the Dodgers before being claimed in the expansion draft for the 1962 season by the newborn New York Mets. In sixteen years, Hodges hit 361 home runs for the Dodgers, batted over .270, and was impeccable with the glove. His brilliant knowledge of the game was confirmed in 1969 when he managed the Mets to a world championship after the club had set standards of futility unmatched since the days of the St. Louis Browns. By 1962, the pieces were in place for five spectacular years of triumph and frustration as keen as any known in Brooklyn. At last freed from the bizarre confines of the Coliseum, the club found Dodger Stadium an ideal home for their exceptional pitching, speed, and timely hitting—qualities that characterize the great team of the 1960s. The stadium opened in April 10, 1962, to a capacity crowd as enthusiastic about the park as those thousands of fans in Brooklyn had been when Ebbets Field opened in 1913. The stadium was "not just any baseball park but the Taj Mahal, the Parthenon, and Westminster Abbey of baseball," wrote Jim Murray in the *Times*.[18] Wally Post's home run marred the inaugural and won the game for the Cincinnati Reds, but many of the honored guests correctly identified the opening of baseball's first modern stadium as the harbinger of a new era.

The 1962 season featured exceptional individual performances. Sandy Koufax began his string of Hall of Fame seasons by becoming the first pitcher to strike out eighteen batters on two occasions when he turned the trick against the Cubs on April 24. He then hurled the first of four career no-hitters on June 30 against the Mets. A circulatory problem known as Reynaud's Phenomenon, about which all Los Angeles baseball fans were to become expert, cut short his remarkable season. Koufax finished at 14–7 with a league-leading earned run average of 2.54. Don Drysdale topped the league with 25 victories and had an e.r.a. of 2.83. Ron Perranoski, mainstay of the bullpen, also topped the league, with 70 appearances. The rest of the staff compiled the balance of the team's 10 wins in the regular season, which had been extended to 162 games in order to make adjustment to the new franchises in New York and Houston.

The Dodgers' attack was spearheaded by Maury Wills, who shattered one of the ancient records of the game by stealing 104 bases, eclipsing Ty Cobb's mark of 96, set in 1915. Tommy Davis won the league batting championship, hitting .346, and he also took the r.b.i. crown with 153, a club record. Frank Howard led the team with 31 home runs in the spacious park. Duke Snider, reduced to a minor role, contributed 5 home runs and 30 r.b.i.'s while batting .278.

In spite of these achievements, the season ended in terrible frustration. The Dodgers lost their final three games and fell into a tie with the Giants, whom they faced in the playoffs with an offense that had been unable to generate a single run in their last two games. The first playoff game was held in the Giants' Stadium, Candlestick Park. Koufax was unable to survive the first inning as the Dodgers were again shut out, 8–0. In the second game at Dodger Stadium, the offense finally erupted for seven runs in the sixth inning and rallied for a win that forced the decisive third game.

On October 4, the National League pennant was finally decided in a manner that brought back the nightmares of 1951. The Giants scored two runs in the third inning when the Dodgers committed three errors, including two throws that hit runners in the back. But the Dodgers fought back on three stolen bases by Wills and a home run by Tommy Davis and took a 4–2 lead into the ninth inning. The frame began with Matty Alou pinch-hitting a single into right-center. Harvey Kuenn grounded into a force play, but the Dodgers' Ed Roebuck, who had pitched seven times in nine games, began to tire and walked the next two hitters, loading the bases. Willie Mays then lined a single off Roebuck's glove bringing home one run and reloading the bases. Stan Williams was brought in to relieve Roebuck, and Orlando Cepeda tied the game with a sacrifice fly to right field. A wild pitch and an intentional walk loaded the bases again with two out. Jim Davenport coaxed a walk on five pitches and put the Giants ahead 5–4. Ron Perranoski was brought in, and he induced Jose Pagan to hit a grounder to second, which Larry Burright booted, allowing Mays to score the fourth run of the inning. The winning pitcher for the Giants was none other than Don Larsen, who had pitched the perfect game against Brooklyn in the 1956 World Series.

Bitter though the Dodgers' defeat was, it is not recalled with the same pangs that surround the 1951 campaign. For one thing, although the Dodgers held a commanding lead in the pennant race, it was nothing like the thirteen games they squandered in 1951; moveover, the stunning victory by the Giants in 1951 reflected on their resilience, while the 1962 playoffs pointed up the inability of the Dodgers to perform basic tasks like making routine throws to the bag and getting the ball over the plate. And the rivalry between Los Angeles and San Francisco is genteel compared with the interborough battles of New York in the 1950s.

Perhaps another reason why 1962 hasn't haunted the team's history in Los Angeles is the terrific rebound the team made in 1963. Before the season began, the final offensive link with the Brooklyn teams was cut when Duke Snider was sold to the New York Mets on April 1. Snider still holds the team record for career home runs, with 381, including a club season high of 43 in 1956. He also drove in more runs—1,271—than any other Dodger. Snider batted .295 in his career, and was arguably as fine a defensive center fielder as ever played the game. For these achievements he was elected to the Hall of Fame in 1980.

Some of the Dodgers' performances fell off a bit in 1963 as the offense generally posted the anemic figures that over the years would become familiar to fans and burdensome to the pitching staff. Wills again led the league in stolen bases, but the total declined from his record 104 to 40. Tommy Davis won another batting title with an average of .326, but his r.b.i.'s dropped to 88. Frank Howard led the team in home runs with 28.

The pitchers responded to the challenge. Blessed with a full season of good health, Koufax won 25 games while posting an incredible e.r.a. of 1.88, both figures leading the league. Drysdale won 19 games and the team itself took 99 to win the pennant by 6 games over the Cardinals. The real drama that season was the World Series against the Yankees. The opening game was held in New York on October 2 with Koufax facing Whitey Ford. Paced by John Roseboro's three-run homer, the Dodgers scored five runs to back Koufax. The brilliant southpaw, who had struck out 306 batters during the season and had pitched his second no-hitter, stifled the Yankees on six hits and struck out a Series record 15.

In the second game Johnny Podres again tormented the Yankees by shutting them out until the ninth inning, when he allowed a single run. The Dodgers secured the victory by scoring twice in the first and adding single runs in the fourth and eighth innings. Bill Skowron, who had been traded by the Yankees, touched his former mates for a solo homer. The final two games were brilliant pitchers' duels. With the scene shifted to Dodger Stadium, Tommy Davis drove in Jim Gilliam in the first inning off Jim Bouton. Thereafter Drysdale and Bouton kept batters at bay inning after inning. In the end, Drysdale made Davis's r.b.i. stand up by shutting out the Yankees on three hits while walking only one, striking out nine.

In the fourth game, Koufax and Ford again faced off and extended the streak of scoreless innings to twelve and a half before Frank Howard homered in the bottom of the fifth and Mickey Mantle tied the game with a solo shot in the seventh. In the bottom of that inning, Jim Gilliam grounded to Clete Boyer at third, and Joe Pepitone lost the throw against a background of shirtsleeves. Gilliam raced to third base, where Willie Davis drove him home with a sacrifice fly. Despite a brief rally by the Yanks in the ninth, Koufax held the lead and the Dodgers won 2–1 on only two hits.

The sweep of the Yankees remains the pinnacle of the Dodgers' history. Having lost six of seven previous Series meetings with the Yankees, the Dodgers still could be motivated by that imbalance; and while many of the stars of the 1940s and 50s had retired, players such as Mantle, Ford, and Podres had played key roles in the games between Brooklyn and New York. The Series demonstrated that, whatever links the Dodgers retained with the Brooklyn clubs, they were a dramatically transformed team. They won the Series with a team batting average of .214, scoring only three runs in their final two games. Maury Wills hit .133 and stole a single base. Tommy Davis hit .400 with two triples, but he had only two r.b.i.'s. While the power hitting of the 1950s had been lost, the pitching rivaled that of any club in the history of the game. Koufax, Drysdale, Podres, and Perranoski compiled an earned run average for the Series of 1.00. The Yankee batting average was .171, and they drew only five walks for the four games while thirty-seven batters struck out. Certainly the final two games were tight contests, which the Yankees could have won to tie the Series. But whatever intangible quality allows one team to gain the critical edge on another favored the Dodgers in October 1963, and a Series that Brooklyn fans must have dreamed of many times was finally realized far from Flatbush.

In 1964 the Dodgers slumped. The pitching staff was hurt when Podres was lost for the year. Koufax won nineteen and led the league with a 1.74 e.r.a., and Drysdale finished 18–16. The balance of the pitching fell to young arms untested in the majors and none of them won more than seven games that year. Wills, and Tommy and Willie Davis had good years, but the team scored fewer runs than any other club except for the expansion Mets and Houston Colt 45's. Wills again led the league with fifty-three steals, but the offense was sufficient to support only brilliant pitching, which the Dodgers could offer only two out of of four outings. The result was an 80–82 record, and a sixth-place finish, tying them with Pittsburgh. The season demonstrated that, although the Dodgers were capable of stifling any team in the majors, they lacked the depth to secure a dynasty. Changes in the club could be expected.

During the winter trades, Frank Howard was swapped to the Washington Senators for Claude Osteen, as the Dodgers sacrificed their one legitimate power hitter for additional pitching. The infield for 1965 was also changed, with the addition of Wes Parker at first base and Jim Lefebvre at second.

When Tommy Davis was lost for the year on May 1 with a broken ankle, the 1965 campaign put an even greater burden on the pitching staff. Lou Johnson, who had been dropped by four other teams, picked up part of the slack along with Lefebvre, who was named Rookie of the Year. Osteen provided fifteen wins but also lost fifteen, despite an impressive earned run average of 2.79. Koufax and Drysdale again anchored the team with tremendous performances, Koufax winning twenty-six games and leading the league again with an e.r.a. of 2.04, while Drysdale won twenty-three and compiled an e.r.a. of 2.77. Koufax's fourth no-hitter in as many years was a perfect game against the Chicago Cubs on September 9. The game was a microcosm of the games played by the Dodgers' team of that era. While the Cubs were thoroughly controlled, the Dodgers threatened to give Koufax a chance at surpassing Harvey Haddix's feat of pitching twelve perfect innings in a losing cause. The only run of the game came on a walk, a stolen base, and two sacrifices, and the Dodgers' lone hit had no bearing on the outcome. Drysdale, who had become accustomed to losing after great efforts, was in Los Angeles in 1964 when Koufax no-hit the Phillies in a road game. Drysdale's reply upon hearing the news: "Who won?"

After a tight race with the Giants, the Dodgers won their third pennant in eight seasons in Los Angeles as Koufax took a 3–1 decision over the Braves on the next to last day of the season. The Dodgers touched Milwaukee for only two hits, both by Lefebvre, but their three runs were more than enough to secure the title.

Los Angeles was favored in the World Series against the Minnesota Twins, who had moved from Washington, D.C., in 1961, leaving behind the sobriquet "Senators." The Twins were a hard-hitting team led by Zoilo Versailles, Tony Oliva, Bob Allison, and future Hall of Famer Harmon Killebrew. The Twins also had formidable pitching in Jim "Mudcat" Grant, Jim Kaat, and Camilo Pascual. The team had won 102 games during the season to finish 7 ahead of the White Sox. The season was the start of a new era for the American League as the Yankees dropped to sixth place, and their dynastic hold on the league, which could be traced back to 1921, was finally broken.

The Twins established their credentials by decisively taking the first two games of the Series in their home park. In the first game, Grant held the Dodgers to two runs, while Drysdale was touched up for six in the third inning. The next day Jim Kaat held the Dodgers to one run, while the Twins got to Koufax for the winning runs in a 5–1 victory. The Series returned to Los Angeles with the Dodgers not only down two games but their Hall of Fame tandem defeated. Claude Osteen came through with a 4–0 shutout over Pascual in the third game, and Drysdale knotted the Series the next day by striking out eleven in a 7–2 win. In the fifth game Koufax returned to form, blanking the Twins 7–0 on four hits, fanning ten.

For the final games the Series returned to Minnesota, and Mudcat Grant took charge in the sixth game, not only limiting the Dodgers to one run on six hits, but hitting a three-run homer to seal the victory and force a seventh game. Manager Walter Alston now faced a tough decision. Drysdale was due to pitch in the rotation, but how could Koufax be left out of a seventh game? Alston concluded that he preferred to start Koufax; he could use the more rested Drysdale in relief if necessary.

With two days rest, Koufax pitched a second consecutive shutout, holding the Twins to three hits while striking out ten. His iron-man performance gave the Dodgers their third world championship in Los Angeles, far eclipsing the entire history of the franchise in Brooklyn. The Dodgers' success threatened to return the game to the pre-Ruthian era of strategic baseball. With such dominant pitching, one or two runs fashioned through walks, stolen bases, and sacrifices, could stand up for an entire game. Even the hard-hitting attacks of the Giants and Braves were frequently reduced to futility when facing the Dodgers' pitching.

Prior to spring training in 1966, Koufax and Drysdale made another contribution to baseball, one that terrorized club officials rather than opposing batters. The two aces formed a mini-union and through an agent asked for three-year no-cut contracts at $166,666 each per year.[19] For the Dodger management, the least of the problems was the money, although the sums represented substantial increases above the $75,000 each was thought to be earning at the time. In the 1960s agents and serious collective bargaining were unheard of. The Dodgers refused to meet the terms, and Buzzie Bavasi, the team's general manager, refused to recognize the right of players to bargain collectively for specific salaries. He also expressed a preference for negotiating directly with the players themselves.

In 1966 salaries in excess of $100,000 were reserved for a few superstars. Later, when players began to earn larger sums, especially with the arrival of free agency and salary arbitration, agents became necessary to structure the payments in the most financially rewarding manner. Compensation got not only larger but also more complex. Koufax and Drysdale's requests and their means for pursuing them were radical. When the Dodgers rejected their terms, the two pitchers exercised the only available option: they refused to report to spring training. This step not only precluded their reaching an agreement with any other team, but the Dodgers could retain rights to the pair in perpetuity. A season-long holdout might not have improved the players' bargaining position.

After missing virtually the entire spring training, Koufax and Drysdale signed on March 30. Koufax was paid $120,000 and Drysdale $105,000. The final agreement was reached with direct negotiations between the players and Bavasi, and the two were paid individually rather than as an entry. "Baseball is an old-fashioned game with old-fashioned traditions," said Walter O'Malley. "If we allowed this entry business to take hold it would lead to practices not possible to tolerate."[20]

O'Malley would live to see the old-fashioned game utterly transformed through the limitations imposed on the reserve clause by arbiter Peter Seiss. That decision was based in part on another salary dispute between Dodger management and one of their ace pitchers, Andy Messersmith. The impact of free agency has led to salaries that would easily have propelled Koufax and Drysdale into the million-dollar-a-year class if they had had the good fortune to be born a few years later.

In 1966, the Dodgers' offensive attack improved somewhat. The team average of .256 matched the league average, and they rose from tenth place in home runs in 1965 to tie for eighth in 1966. But pitching was the story again, as Koufax won twenty-seven games with another championship e.r.a.—1.73. Drysdale fell off to 13–16, but Osteen won seventeen, and Phil Regan became the new ace of the bullpen with a 14–1 record, a 1.62 e.r.a., and a league-leading twenty-one saves. The season was one of streaks. In April, the Dodgers were shut out 2–0 on three successive nights by Ferguson Jenkins, Ken Holtzman, and Larry Jastrow, all rookies. Jastrow went on to torment the Dodgers all year, shutting them out four consecutive outings. In September the Dodgers staff put together its own shutout streak, blanking opponents for forty straight innings. The pennant race was another struggle between the Dodgers and the Giants. Sandy Koufax again pitched the decisive game, a 6–3 win over the Phillies in the final game of the season, holding off the Giants, who finished only a game and a half back. The title was the club's third in four years, an achievement akin to that of the Boys of Summer in the 1950s. The Los Angeles team had the added distinction of winning World Series titles which had all but eluded Brooklyn.

In 1966 the Dodgers faced the Baltimore Orioles—the reconstituted St. Louis Browns—winners of their first pennant on the East Coast. The Orioles had been bolstered by the acquisition of Frank Robinson from Cincinnati. Robinson promptly won the Triple Crown, topping the American League in batting average, home runs, and runs batted in. Another Oriole, Brooks Robinson, was causing people to question if Pie Traynor should still be considered the best third baseman of all time. First baseman Boog Powell added more power to the Orioles' attack, and Luis Aparicio, picked up from Chicago, secured the middle of the infield. The pitching staff appeared adequate. Their leader was rookie Jim Palmer, who had won fifteen games. Dave McNally, who would achieve lasting fame through an arbiter's decision some years later, had won thirteen. Wally Bunker and Steve Barber each won ten, respectable numbers but hardly likely to make anyone forget Koufax and Drysdale.

The Series shaped up as something like the preceding year's—Dodger pitching against a power-hitting club. But the outcome was considerably different. The Series opened on October 6 in Los Angeles with Don Drysdale facing Dave McNally. The Orioles broke on top with successive home runs by Frank and Brooks Robinson in the first inning. The Dodgers' offense mustered six hits, all singles as they were shut out brought in and stifled the Dodgers completely, striking out eleven, including six in a row at one point, as the Orioles won 5–2. In the second game the Dodgers' defense collapsed behind Sandy Koufax. Willie Davis made three errors in the fifth inning, two on the same play. His mates added another three errors to tie the Series record for bungling. Jim Palmer held the Dodgers to four hits and, nine days shy of his twenty-first birthday, became the youngest pitcher in World Series history to hurl a shutout. The quick turnover of baseball generations was evident in this game, since Palmer used to watch Koufax pitch in the Coliseum when the Oriole pitcher was a high school student in Los Angeles.

The Dodgers were down two games to none, as they had been the year before. Against Minnesota, however, they returned to Dodger Stadium for the next three games. They now faced the task of going to Baltimore to catch up. Claude Osteen limited the Orioles to three hits in the third game, but one of them was a home run by Paul Blair. The Dodgers' offense mustered six his, all singles as they were shut out once again, this time by Wally Bunker. Don Drysdale took the mound to avert a sweep, and he, too, pitched well. The Orioles managed only four hits, but again one of them was a homer, by Frank Robinson. The Dodgers' offense failed to score again for a total of thirty-three consecutive innings without crossing the plate, too anemic a showing even for Koufax and Drysdale to rescue them.

Although the outcomes were completely reversed, the 1966 Series resembled that of 1963. The final two games of each were razor-thin contests that could easily have gone the other way. It is tempting to think of the 1963 sweep as the accomplishment of a great team while the team swept in 1966 was an assemblage of pretenders, but fundamentally the same club was involved in both contests. Dodger pitching, after all, held the powerful Baltimore attack to an average of .200 and allowed only 2.65 earned runs per game. Once again the pitching was sufficient to keep the Dodgers in the games, and to give the offense a chance to produce the victory. In 1963, three runs were enough to win the final two games. In 1966, four runs would have turned the trick, but the offense fell into the kind of slump that can infect any team. Only terrible timing obscures the achievement of the 1966 team.

Following the season, some changes took place. First Jim Gilliam announced his retirement. Baltimore was an appropriate place for him to end his career, since he had begun playing for the Baltimore Elite Giants in the old Negro League before Robinson broke baseball's color barrier. Although Gilliam achieved prominence in Los Angeles, he was another link to the Brooklyn era, and his departure marked a further transition in the Dodgers. A post-mortem in the *Times* speculated about the Dodgers' future. While noting the Dodgers' pathetic offense, the writer raised this question: "Where would the Dodgers be in a tough league if they didn't have those 27 wins by Sandy Koufax?"[21] Everyone would find out soon enough, for on November 18, 1966, Sandy Koufax announced his retirement. At the age of thirty-one, he seemed to have several more twenty-game seasons ahead of him. Indeed he was at the peak of his career. But the strain of pitching had caused arthritis to develop in his left elbow, a condition that had caused him to throw in pain for several seasons. Rather than risk a severe and permanent impairment, Koufax chose to end his career. In his first six years with the Dodgers, he won a total of 36 games. In his final six, he was the premier pitcher in the major leagues, amassing 129 victories. He led the National League in earned run average each of his last five seasons; led the league in strikeouts four times; and he added four no-hitters into the bargain. He also won three Cy Young awards in a time when only one was allotted for both leagues. Koufax was not as durable as Warren Spahn or Don Drysdale, but in each of his final two seasons he pitched more innings than any other pitcher in the league. He also worked with little rest when pennants or world championships were on the line, and on those occasions recorded some of his best performances.

In 1971, his first year of eligibility, Koufax was inducted into the Hall of Fame. He can legitimately be considered among the very best pitchers of all time and he certainly remains the preeminent pitcher in the Dodgers' history. His development in the 1960s was the clearest distinguishing feature of the Dodgers team that emerged in Los Angeles. He drew many new fans to the Dodgers, and provided Los Angeles with a rich baseball lore before the Dodgers had even celebrated their tenth anniversary on the West Cost. In a community that manufactures celebrities, Koufax displayed gentlemanly reserve, and his achievements in Dodger Stadium erased many memories of the bitter contest over the Dodgers' move to Los Angeles and the stadium controversy that followed.

Koufax's retirement ended the reign of the one great team that has performed in Los Angeles. In 1967, the club collapsed to eighth place with a 73–89 mark and the following year they finished at 76–86, tied for seventh with the Phillies. The 1968 season included a record for consecutive scoreless innings pitched by Don Drysdale, who had labored somewhat in Koufax's shadow. For fifty-six and a third innings in May and June, Drysdale held opponents without a run. In a grim night, Drysdale broke the National League record for consecutive scoreless innings in Dodger Stadium against the Pirates, and received congratulations from Robert Kennedy only hours before the senator was slain on the night of his own victory in the California primary.

In 1969 the Dodgers rebounded to a winning record. Claude Osteen and Bill Singer each posted twenty wins, the last time that feat has been achieved by Dodger pitchers. That year was the first of divisional play, an accommodation to scheduling induced by the addition of franchises in San Diego and Montreal. The Dodgers won eighty-five games that year, good for a fourth-place finish in the six-team National League West.

Drysdale, a victim of a torn rotator cuff, accounted for only five wins as his career ended at the age of thirty-three. He was the last Dodger to have played in Brooklyn, where he posted a 5–5 record in his rookie year of 1956. He made a brief appearance in the World Series that year. In the final season at Ebbets Field he was 17–9, with a 2.69 e.r.a. His career had been somewhat stormy in Los Angeles, where he was expected to develop immediately as the ace of the staff. Koufax's emergence helped reduce those expectations to more realistic levels, and the move to Dodger Stadium saw Drysdale respond with a twenty-five-win season in 1962. His reputation grew in tandem with Koufax's, although Drysdale was more notorious because of his willingness to fire an occasional fastball at the ribs of a hitter who leaned too close to the inside part of the strike zone. At the end of his career, Drysdale had won 209 games and had career e.r.a. of 2.95. He also hit 29 home runs, Ruthian numbers for a pitcher. In 1984, Drysdale joined Koufax in the Hall of Fame, the second Los Angeles Dodger to claim that honor. While Koufax had been born and raised in Brooklyn, Drysdale's childhood was spent in Los Angeles. In the 1970s the Dodgers developed another outstanding team, but it remained a stride behind the great teams of that era in Cincinnati and Oakland. Don Sutton, Andy Messersmith, and Tommy John anchored the pitching staff, and they looked for support to a parade of players that included Ted Sizemore, the 1969 Rookie of the Year, Bill Grabarkewitz, Willie Crawford, and Bill Sudakis. In 1971 Bill Buckner and Steve Garvey joined the Dodgers, and began to supply the offense that had been missing since the final years in Ebbets Field. That year the Dodgers missed a division title by only one game, as the Giants edged them out.

The following year Cincinnati's Big Red Machine won the National League title and became the first of three teams to lose the World Series to the Oakland A's. The Reds were led by Pete Rose, Johnny Bench, and Joe Morgan, whose hitting and defensive skills carried an average pitching staff, reversing the pattern the Dodgers had established in the 1960s. As great as the Reds were, Oakland had Reggie Jackson, Sal Bando, Joe Rudi and Gene Tenance, who supported a pitching staff that starred Jim "Catfish" Hunter, Vida Blue, Ken Holtzman, Blue Moon Odom, and Rollie Fingers. Against such competition, the Dodgers were relegated to the role of bridesmaids.

After losing a ten-game lead over the Reds in 1973, the Dodgers rebounded to capture a National League pennant in 1974. Garvey became a fixture at first base, setting a league record by playing 1,207 consecutive games. He was joined by Davey Lopes at second base, Bill Russell at short, and Ron Cey at third base, an infield that eventually played together longer than any other in the history of the game. Jim Wynn played centerfield, Buckner moved to left, and Willie Crawford was in right. Steve Yeager provided the first stability at catcher since John Roseboro had retired. This team, though superior to that of the 1960s offensively, resembled the 1950s Dodgers in generally coming up a dollar short against the one or two better teams of that time. In the 1974 Series, for example, Oakland dispatched the Dodgers in five games, four of those games, including the Dodgers' only victory, were decided by 3–2 scores. A very good team had lost to a great one.

Cincinnati reasserted itself in 1975, winning 108 games to finish an incredible 20 ahead of the second-place Dodgers. The margin was cut to 10 in 1976, with Cincinnati winning its second pennant in a row and third in five years. The Dodgers had strengthened their outfield by acquiring Reggie Smith and Dusty Baker through trades, but the gap between the Dodgers and Reds was not to be closed through the addition of a couple of quality players.

At the end of 1976 season, Walter Alson announced his retirement after twenty-two years of managing. His 2,040 wins made the fifth highest total of any manager in history and more than any other Dodger pilot. He had also guided the team to seven pennants and four world championships. Under his direction the Dodgers had also finished second eight times and third twice. In many of those years, Alston had stoically endured speculation that he would be dismissed at season's end in favor of a more boisterous character, whom some members of the press assumed would win more pennants. But Alston won world championships in three different home parks and with markedly different types of teams. And he asserted his authority unmistakably whenever his youthful troops mistook his calm demeanor for ineptness. In 1983, he was elected to the Hall of Fame, one of only a handful chosen solely for their managerial skills.

Tom Lasorda, Alston's successor, brought an ebullient personality to the job that fit the expectations of those in the media who gravitate toward celebrities.[22] Frank Sinatra, Don Rickles, and other entertainers whom Lasorda numbered among his personal friends occasionally visited the Dodger clubhouse leaving behind autographed pictures. The alliance between Lasorda and Hollywood celebrities further secured the Dodgers' place in the entertainment community. The team receives coverage outside the sports pages, which enhances its appeal to the community and attracts even more fans to the ballpark. Indeed Lasorda has become something of a national celebrity whose gregarious manner and penchant for quotable remarks have blinded some people to his managerial skills. In his first ten years at the Dodgers' helm, the team has won five division titles, three pennants, and a world championship.

He has won, moreover, with a team that he himself nurtured. Garvey, Russell, Cey, Lopes, and other Dodgers stars of the 1970s received their introduction to professional baseball from Lasorda on Double A and Triple A minor league teams. Even after the facade of a happy family no longer masked the personal strains on the team, Lasorda was still able to run the club successfully. The only world championship the Dodgers have won since 1965 came in 1981 when Lasorda not only overcame the disruption of an extended player strike, but also guided the team to dramatic rallies in the National League playoffs and the World Series.

When the stars of the 1970s began to show signs of having passed their peaks, they were pushed out by another bounty of talented young minor league players who surfaced in Albuquerque. The prospect of getting similar production for lower salaries was a factor in the Dodgers' deciding to part with players like Garvey, Lopes, Cey, and Tommy John. With a new cast of unproven players like Greg Brock, Mike Marshall, and Pedro Guerrero, Lasorda again guided the team to a divisional title in 1983.

The success of the current team is based on one of the best pitching staffs in baseball, led by two young stars, Fernando Valenzuela and Orel Hershiser. The offense can be spotty, but in Guerrero and Marshall the team has two of the best young power hitters in the league. Defensive lapses, especially Steve Sax's throws from second base, occasionally sink the team to comic lows, but it is a typical Los Angeles Dodgers club. They are one of the best teams in baseball, and if the breaks go their way they can win a world championship; but at this point they are not in a position to dominate as the team of the early 1960s did.

The organization's continued ability to field competitive teams is all the more remarkable when one notes the fundamental changes in baseball during the past decade, particularly free agency, which has made it possible for teams to pursue championships through free spending on salaries subsidized by other corporate assets. Unlike other teams, the Dodgers' revenues for player development, salaries, and free agency must be generated by baseball earnings rather than through shipping, hamburgers, or broadcasting. The Dodgers' principal venture into free agency left them with staggering contracts for two pitchers unable to perform up to expectation. Since then, the organization has withdrawn from the competition for free agents, and relies almost exclusively on its farm system.

The new realities of the contemporary era have been joined by the darker problems plaguing baseball, most notoriously drug use. A dramatic case involved Steve Howe, who had been the league's Rookie of the Year in 1980. After he won a game against the Yankees in the 1981 World Series and saved the final game, Howe appeared to be in line to succeed Sherry and Perranoski in the tradition of star relief pitchers. But recurrent cocaine use changed Howe into an erratic figure who became too unreliable for any team, despite his estimable gifts. On the other hand, Bob Welch has set a courageous example in controlling the alcoholism that threatened his career. Throughout the history of baseball, an adolescent part of the game's romance has included the exploits of players who performed through the haze of alcohol or after all-night amorous adventures. Drugs depart from that tradition because they involve players in illegal activities, perhaps in organized crime. The prospect of a player incurring obligations to people who might profit from fixing games threatens to return the sport to its seamy early decades when the integrity of the game was in question. Since the Black Sox scandal was dealt with in 1920, baseball has been free of

pervasive suspicion about the legitimacy of the sport itself. Now baseball finds it must confirm its integrity once again.

Any discussion of the close ties between the Dodgers and their fans
in Los Angeles must mention the team's principal announcer, Vin Scully.
In recent years, Scully has become familiar nationally through his network broadcasts of baseball, football, and golf. In addition to those
responsibilities, Scully continues to broadcast the Dodgers' home games.
Despite his youthful appearance, Scully represents another link to the
Brooklyn franchise. He began to announce Dodger games after graduating from Fordham in 1950. He thus has the perspective of someone
close to the team when Walter O'Malley wrested control from Branch
Rickey, when Jackie Robinson was in the early part of his career, when
the unknown Walter Alston was tapped for the manager's job, when
Bobby Thomson hit the famous home run, when Johnny Podres pitched
the Dodgers to their first world championship, and when the Dodgers
played their final season in Brooklyn.

Perhaps more than any player, Scully contributed to the Dodgers'
appeal in Los Angeles. He tempers his enthusiasm to give an honest
account of the action before him, and he never loses his awareness that
in the end he is describing a little boy's game. The development of
transistor radios in the late 1950s enabled fans to bring radios to the
game so that they could continue to enjoy Scully's presentation, a practice not always understood by writers. The explanation is simple, however: prior to 1958 the technology did not exist, and since then few
other announcers have been worth listening to. Scully himself says that
the phenomenon of transistor radios in the ballpark originated in the
days of the Coliseum, where fans sat so far from the field that they
brought radios to help them follow the action.[23] When Dodger Stadium
opened, the habit had become ingrained, and continued. Not only is
the transistor used more prevalently in Dodger Stadium than in the
other parks Scully visits in his network broadcasts; he reports that their
use in Los Angeles occasionally causes technical feedback problems
through the crowd microphone. Scully's popularity affords him the opportunity to influence the way fans perceive the game, especially in its
larger economic and social contexts. His upbeat but realistic approach
emphasizes positive aspects of the game and of the players and he
avoids any temptation to preach about their ills.

From the romantic perspective of baseball, Los Angeles has developed a rich tradition. The years in the West have included thrilling
victories and championships as well as heartbreaking defeats. The
Dodgers have fielded a few players of Hall of Fame caliber and other,
minor stars who have endeared themselves to fans despite brief or sta-

tistically unimpressive careers. Al Ferrara, Jay Johnstone, Wes Parker, Manny Mota, Jim Brewer, Art Fowler, Dick Nen, and Bill Singer are among the players who have generated excitement for the millions who have followed the team since 1958.

In diverse ways, the Dodgers have developed ties in Los Angeles. Personnel decisions in response to free agency have triggered passionate reactions from the many fans of Steve Garvey, Andy Messersmith, and Tommy John. The decision to go with young players has led popular players such as Dusty Baker, Bill Buckner, and Ron Cey to be traded. The Hispanic community in Los Angeles has been gratified by the emergence of stars like Valenzuela, Guerrero, Alejandro Peña, and Mariano Duncan. Families who have struggled with the tragedy of drug abuse have seen that wealth and fame are not barriers to that lamentable illness. The heroic struggles of Tommy John and Sandy Koufax against serious injuries have set another kind of example. Finally, the game itself, the duel of pitcher and hitter, the strategic maneuvers of competing managers, the brilliant feats of world-class athletes, as well as the dropped fly balls, have enthralled the community. As the *Times* had anticipated, a diverse and parochial group has from time to time cohered around a common interest in the fate of a baseball team.

The history of the Dodgers in both Brooklyn and Los Angeles demonstrates that the romance of baseball is universal. The emotional attachments a community may feel for a team are insufficient to ensure that the franchise will remain where it is beloved. Major league teams will always be a scarce commodity, and cities will inevitably compete for a limited number of teams. Disappointment and even bitterness will accompany the allocation of franchises.

The business and romantic perspectives provide compelling, albeit conflicting, answers to whether the Dodgers should have moved to Los Angeles. A final perspective remains more significant, in part because it is created by the clash of the first two and also because this *civic* perspective is the arena in which public choice resolves the competing claims about franchises and their communities. This perspective is especially important for other cities facing issues similar to those raised by the Dodgers in the 1950s.

The Dodgers were not the first baseball team to change cities, but they were the first to do so with great controversy. They have not been the last. Seattle, Milwaukee, and Washington, D.C., have all lost baseball franchises amid allegations that legal and moral bonds to the community were improperly severed. Denver, New Orleans, and Phoenix have debated the propriety of spending public money to attract a franchise that would enhance the status of the town as well as stimulate local business. Chicago is currently struggling to keep both the White Sox and the Cubs.

In the Dodgers' move, the civic perspective remains controversial but the dimensions of the arguments at least are clear. They are also instructive. Rather than consider franchise relocation and new stadiums on an ad hoc basis, other cities should develop a framework for analyzing issues and making use of the lessons learned in previous cases. This framework should identify three basic models local governments can employ when a team alleges a need for a new stadium and threatens relocation if the need is not met.

The first of these models is the classic free market economy, which states that government has no role to play beyond the protection of private property, and that no aid should be extended to any business in its efforts to secure land and facilities.[24] The concept of the free market has always claimed a powerful hold on American culture, but from our nation's inception government has actively promoted certain businesses through tariffs, land grants, the awarding of monopolies, municipal bonds, and other measures. This history of market intervention shouldn't obscure the point that government aid to baseball teams is not essential and should occur only when various community representatives have determined that the public interest is served by such action.

A second basic model in this civic perspective considers sports franchises and stadiums as a public interest to which the government should extend some limited support. From this viewpoint, the free market does not cover the wide range of public interests; highways, schools, and police departments, for example, can only be secured through the collective action of a community. A free-market model requires that a baseball franchise purchase from each current owner the land needed for a new stadium, and then build its own facility. But even in the early part of the century, when the section of Brooklyn that he pursued was relatively undeveloped, Charles Ebbets had trouble employing that technique. In the modern age, no team could contemplate purchasing land in such a manner. The individual parcels would steadily increase in value until the final one potentially cost as much as all the preceding plots combined, as nearly happened with the final twelve plots that the Dodgers purchased in Chavez Ravine. A public-interest model would recognize a role for the state in assembling the land for sale to the team through a device like the Federal Housing Act or through eminent domain. The public-interest model is intended to adapt the market to the social and economic conditions that have developed since Adam Smith proposed free market theory in 1776. A good illustration of this model is Los Angeles's response to the Dodgers' interest in a stadium. Dodger Stadium was privately built and operated, but the participation of government was indispensable in providing the opportunity for that private venture.

The third model in this civic perspective has characterized the construction of virtually every modern sports stadium, except for Dodger Stadium and a few other facilities. The model, is, in a word, socialism, with the government constructing the facility as a public venture financed through municipal bonds or tax money. While socialism remains abhorrent to most Americans, many of our public enterprises—transportation, recreation, and utilities—are owned by the public, operated by government agencies, and thus obey the principles of socialism. Critics point to the postal system and Amtrak as examples of the inefficiencies of such structures, but they forget government's role in directing the interstate highway program, the national parks, and the Apollo space program.

The Socialist option was exercised by New York when it faced losing the Yankees, the third case of an established New York team playing in a decaying stadium in a deteriorating neighborhood. During the administrations of Mayor John Lindsay in 1965–1973, the Yankees voiced the same complaints expressed by Walter O'Malley and Horace Stoneham a decade earlier. Threatened with losing the team that had brought so much fame to New York, and keeping in mind the city's experience with the Dodgers and Giants, the Lindsay administration intervened more directly than had the Wagner administration. More than $100 million was committed to renovation of Yankee Stadium bringing the aging park up to modern standards of comfort and convenience.[25] Title to the stadium was transferred to the city, which then leased the stadium back to the Yankees. The team, of course, remained in the South Bronx, and rebounded to glory by capturing two world championships in the 1970s, both against the Dodgers.

Given the realities of modern sports, the free market option is not a serious one. Business and romantic attachments to major league franchises both compel some kind of public effort to accommodate the demands of a team in need of a modern stadium. The major issues are determining the degree to which government should be involved, deciding who should bear the financial burdens, and assessing the impact of new construction on established businesses and residents. Controversy arises, in other words, between the public interest and the Socialist models.

Several points should be considered in comparing these options. First, because sports enterprise remains a frivolous undertaking—children's games played for spectators' amusement—one can argue that the state should be minimally involved in keeping the team, even if the community favors retention. Schools and hospitals should not be forced to compete with private entertainment firms for claims to the public treasury, no matter what financial benefits are returned to the community by its teams. Public investment should be as limited as possible, and government revenues should come in the form of taxes rather than rent.

Comparing these models raises a second point: franchise relocation. A community that adheres to limited support for a franchise faces the possibility that owners of a cherished team may opt for a more commodious setting. But this prospect is more remote than is often realized—no baseball team has moved since 1972, when the expansion Washington Senators shifted to Arlington, Texas. The possibility of further moves can't be dismissed; weak ownership and demographic changes will always threaten marginal franchises. How communities can responsibly retain a hold on their teams remains a vital issue.

In recent years, while baseball has enjoyed some calm, other professional team sports have seen a number of franchises relocate, to the extent that the business structures of football, basketball, and hockey have been challenged.[26] For now baseball's continued exemption from antitrust regulations gives its leagues more control over individual clubs, but that privilege is a basic defect that at some future point will be redressed, and the leagues as well as the affected communities ought to have a policy in place to deal with the changed situation. For now the issue has been restricted to the federal level: Congress has considered legislation restricting the movement of clubs.[27] This legislation is problematic because it is based on the dubious antitrust exemption, and because it simply perpetuates the delusion that federal, rather than local, government is the appropriate place to halt the movement of franchises.

Since franchises relocate for financial reasons, a device that addresses this issue directly is a more effective solution to unwarranted shifts than legislation that simply forbids them. One solution is for communities not to construct municipal stadiums, thereby forcing club owners to finance them privately. A number of devices exist to provide such financing, including venture capitalism, equity offerings, and limited partnerships. If the financing is structured in a way that gives the investors part-ownership in the franchise, the expenditure for the stadium will tie the team to the community more effectively than a law restricting franchise movement. Owners who wanted to move their franchises would have to sell the stadium, which would be of interest only to another ball club. Dissatisfied owners would probably sell the team rather than move it.

An incidental benefit of private financing is that investors might prefer stadiums that are smaller and less costly. Modern ballparks resembling Wrigley field or Fenway Park could replace multipurpose monstrosities like Minneapolis's Metrodome. In smaller settings, baseball could recover its more appropriate function as entertainment, and its financial base could be strengthened by owners who are serious about the game and disciplined by market forces.

If practiced in only a few cities or suburbs, this public good model has profound limitations. Rather than submit to it, owners may well prefer to move their teams to cities where the Socialist option is available. The check on this activity lies in the prudent exercise of citizenship: municipal bonds for the purpose of stadium construction must be rejected. The recent tax law may prove to limit the attractiveness of municipal bonds and prevent their use as subsidies for lucrative businesses.[28] When thoughtful observers such as Howard Cosell argue that franchise relocation may arguably be legal but clearly is immoral, they are focusing on a secondary issue.[29] If modern stadiums and arenas were privately financed, the team would be tied to the community in a way that would preclude frivolous shifts from city to city. What is immoral is not the casual transfer of sports teams but the expenditure of hundreds of millions of public dollars for private entertainment businesses. The argument that the community benefits from stadium rental ignores the obvious source of revenues available through taxation on a privately owned stadium.

In the past, cities have preyed on one another by luring existing franchises to replace a departing team. This unseemly and irresponsible use of resources often subordinates the public good to the narrow interests of large corporations for which the team may be a mere marketing trinket. The several congressional hearings on franchise relocation in the 1980s and the litigation that followed the Oakland Raiders' move to Los Angeles may well be obviated by more responsible behavior at the local level. The municipal stadium is a most pernicious form of welfare awarded to millionaire sports owners.[30] Perhaps now that a sufficient number of cities have been burned by franchise relocation, they will discover that a saner public policy is to tell teams they are welcome only if they build their own stadiums.

City officials in New York, commonly seen as falling short of Walter O'Malley's demands, actually went too far. By declining to assemble

the land at Atlantic and Flatbush, which O'Malley could have purchased, New York bypassed an appropriate government response. By
proposing instead a public stadium in either Brooklyn or Flushing, the
city went overboard in committing public resources to private entertainment. A report from New York City's Office of the Comptroller on
the impact of professional sports on the local economy began with this
statement: "Increasingly, owners of sports franchises demonstrate their
portability. This has impelled the City to commit millions to keeping
or attracting professional sports teams in and to New York."[31] This
bureaucratic syntax conceals a remarkably wrongheaded idea, that the
great cities of the country must dance to a tune played by a handful of
wealthy owners who unabashedly pursue welfare in amounts exceeding $100 million during a time of widespread fiscal austerity.

In Los Angeles, the electorate was indisposed to use public funds for
a stadium, and even scrutinized the private purchase of Chavez Ravine
with extraordinary care. The city and county governments played the
limited role of extending for sale to the Dodgers the land which the
city had owned at Chavez Ravine. Since then the public treasury has
received tens of millions of tax dollars from the Dodgers while limiting
public investment to grading and road construction.

In both New York and Los Angeles, public officals appear to have
acted in good faith, promoting what they believed were the best interests of their constituents. Brooklyn was disadvantaged because its voice
was lost in the federated Board of Estimate, where each borough en-
joys equal power, an arrangement recently repudiated as unconstitu-
tional for denying equal representation to the residents of the more
populous boroughs, such as Brooklyn.[32] Los Angeles, even more dis-
persed geographically, has a governmental structure that represents the
interest of local communities less rigidly. In a sense Los Angeles made
less of a commitment to the Dodgers than New York was prepared to
do, but the nature of the commitment was more appropriate to Walter
O'Malley's interests. The perception that O'Malley played New York
and Los Angeles against each other is at odds with the facts; the two
cities were motivated by internal concerns rather than by a competitive
desire to outbid one another for the prize of the Dodgers.

O'Malley was unquestionably a shrewd businessman unaffected by
sentiment in his operation of the Dodgers, but in the end he lacked
the influence in New York to exert his will. In Los Angeles, O'Malley
found that his interests fitted those of the city, which desired a major
league baseball team and wanted also to dispose of the land at Chavez
Ravine. O'Malley's business skills deserted him in his estimation of the
risks awaiting him in Los Angeles, but his luck prevailed as the elec-

torate rejected Proposition B and the state supreme court reversed Judge Praeger's repudiation of the Chavez Ravine contract.

From the civic perspective, the Dodgers had received no special benefits from New York as of 1957. They left neither a stadium whose site had been assembled by governmental action, nor a municipal stadium constructed at public expense. Ebbets Field had been privately financed and operated for over forty years. The free market model was in operation, and, according to it, O'Malley had as much license to move his business as did any other entrepreneur. The perception of Walter O'Malley as a villain not only evades the real causes of the Dodgers' move; it also obscures the achievements of the franchise in Los Angeles for which O'Malley merits inclusion in the Hall of Fame, joining his predecessors, MacPhail and Rickey.

The fact that the Dodgers' move can be justified from the civic perspective is, of course, no consolation to the fans in Brooklyn. The lesson for cities with similar emotional bonds to a franchise is that those feelings must be translated into commercial or financial ties that can't be broken so easily.

In the years after the Dodgers' move, Brooklyn and New York City encountered some of the grimmest times in their history.[33] Race riots produced the worst violence in the city since the Irish rebelled against military conscription during the Civil War. Cohesion and civility ceased to govern the institutions of the city during the 1960s as unions, universities, banks, corporations, and politicians singled one another out as scapegoats for New York's vast social and economic problems. The city that thought itself wealthy enough to build Shea Stadium and renovate Yankee Stadium was too poor a few years later to pay its teachers.

During this time, Brooklyn's suffering was particularly acute. Once the borough had proudly argued that it had the best team in the country; now it was reduced to bitter debates about whether it contained the nation's worst slum, Bedford-Stuyvesant. To the biblical plagues of hunger, disease, and ignorance were add new scourges, such as drug addiction. Recently, the city has turned around somewhat. The solution had nothing to do with baseball; the Mets' miracle was confined to the diamond, and the Yankees remained the only visible success in the South Bronx. The reversal was achieved, rather, through responsible decisions made by those with authority in government, finance, labor, and other vital institutions; and while many thousands still are disadvantaged, the city can justly point to improved conditions for millions of people and businesses.

Brooklyn has also begun to recover and reassert its own unique culture. The Atlantic-Flatbush neighborhood, which so easily could have been the Dodgers' new home, is at last the object of a redevelopment project. The Dodgers were turned down, but new opportunities are being given to others to develop businesses and provide jobs in a crucial part of the borough. The residential neighborhoods of Brooklyn occasionally are featured in the *New York Times* real estate section, which calls attention to and promotes the gentrification of old neighborhoods. The new class of wealthy young inhabitants introduces different challenges, but they are certainly preferable to the old blights of neglect and decay.

A final sign of Brooklyn's return is the effort to restore professional baseball to the borough. The fanciful appeals to the Dodgers to return have been replaced by realistic attempts to locate a Triple A franchise near Coney Island.[34] For the moment these efforts are stalled because the New York teams may invoke a territorial prerogative forestalling the potential competition. If that shortsighted maneuver is reversed, Brooklyn may again develop baseball as a vital part of its culture. A top minor league team is a long way from the Boys of Summer, but it is also a stride forward to a realistic and productive future and away from the mire of a bitterly nostalgic past.

During its years as home to the Dodgers, Los Angeles has faced many of the same problems confronting all great American metropolitan areas. The 1960s outbreak of race riots began in Watts in the summer of 1964. Gang wars have introduced to some barrios a measure of the terror that grips Beirut. Drugs and illiteracy thrive alongside fashion and wealth.

At the same time the city boasts achievements that would make any community proud. In 1973 Los Angeles became the first city in the country with a minority black population to elect a black mayor. Tom Bradley has been reelected three times, and narrowly missed winning the governership in 1982. Faced with the challenge that New York met a century ago—that of being our major urban center of assimilation and opportunity as immigrants have flooded the United States from Central and South America and from Asia—Los Angeles has performed the role with remarkable success, providing children from utterly diverse backgrounds with the tools for economic and social advancement.

In 1984 the city mingled Hollywood glitter, administrative efficiency, and fiscal prudence in hosting the Summer Olympic Games.[35] Officials were financially restrained by the same public scrutiny that so earnestly reviewed the Dodgers' case. Few major facilities were constructed for the games—the Coliseum was upgraded to function as an

Olympic stadium, just as it had been in 1932. The growth of Los Angeles sports since 1958 was evident in the availability of other sites for contests, such as the publicly owned and little-used Sports Arena and the Forum in Inglewood, the home of basketball's Lakers and hockey's Kings. Los Angeles even introduced baseball to the Olympics as a demonstration sport. Those games were held, fittingly, in Dodger stadium, returning the favor owed since 1958, when the Olympic Stadium had accommodated the Dodgers upon their arrival from Brooklyn.

Demographic projections indicate that by the turn of the century Los Angeles will have replaced New York as the most populous city in the United States.[36] The reversal of status will introduce new challenges to both communities. For New York the era of cheap labor migrating from Europe, which helped sustain the industrial age, will have passed and the city will have to find new ways to serve its citizens. Los Angeles will have to meet new demands for services in a city even more diffuse and varied than it is today. To that end, the role of baseball is unmistakably minor. From time to time, however, the diverse population that constitutes the nation's largest city will set aside its immediate and pressing concerns and become lost for a moment in the adventures of the Los Angeles Dodgers. Few will consider the anomaly of the name "Dodgers" in a city where automotive courtesy is prized among the highest civic virtues, perhaps fewer still will reflect on a place once known as Pigtown, where the team had its birth. They will simply be drawn to the excitement of a child's game being played in a magnificent stadium, and they will enjoy, at least temporarily, a sense of community.

The Chavez Ravine Agreement

A G R E E M E N T

THIS AGREEMENT, made and entered into this 3rd day of June, 1959, by and between THE CITY OF LOS ANGELES, a municipal corporation (hereinafter called City), and the LOS ANGELES DODGERS, INC. (formerly known as the BROOKLYN NATIONAL LEAGUE BASEBALL CLUB, INC.), a New York corporation (hereinafter called the Ball Club),

W I T N E S S E T H :

WHEREAS, City is the owner of certain property in an area generally known as Chavez Ravine, and is in the process of acquiring additional property in said area; and

WHEREAS, such property is no longer required for the use of the City; and

WHEREAS, the bringing of major league baseball to Los Angeles would result in direct and indirect benefits to the City, and would be highly beneficial to the City, to the public, and to its inhabitants to have such property the site of a major league baseball stadium; and

WHEREAS, the placing of such property on the tax rolls would produce substantial additional property tax revenues to the City; and

WHEREAS, the City would receive substantial tax revenues from sources, other than the Ball Club, incident to major league baseball in Los Angeles; and

WHEREAS, the Ball Club is willing to acquire such property, and would, at its cost and expense, construct a major league baseball stadium for the purpose of providing facilities for the major league baseball club known as the LOS ANGELES DODGERS (formerly known as the BROOKLYN DODG-ERS); and

WHEREAS, such Club has a long-standing policy of admitting juveniles to various games free of charge as a method of combating juvenile delinquency and stimulating interest in healthful recreational activities; and

WHEREAS, as part of this policy said Ball Club will construct recreational facilities costing not to exceed a half million dollars on a 40-acre portion of the property to be conveyed by the City, as herein provided, in Chavez Ravine, such recreational facilities to be mutually agreed upon by the parties hereto before such property is conveyed to said Ball Club, such recreational facilities to be constructed simultaneously with the construction of said stadium; and as a further part of said policy Ball Club will maintain such recreational facilities for a period of 20 years, at an annual cost of $60,000, but shall be under no obligation to furnish personnel for the operation thereof; provided, that in the event such maintenance cost in any one year does not amount to $60,000, then as a further part of said Ball Club policy the difference between such maintenance cost and $60,000 shall be paid to City so that it may be used by City for the providing of or maintaining recreational facilities elsewhere in said city.

In this connection, it is understood that such recreational facilities shall be under the control of City, but City shall not permit the scheduling of any event at any of the recreational facilities on said 40 acres which would involve the concentration of a large number of participants or spectators during any time when Ball Club has scheduled an event at said baseball stadium, and that during any event at said baseball stadium Ball Club shall have the right to use any facilities suitable for parking as may be provided within said 40 acres; and

WHEREAS, such additional recreational facilities are sorely needed by the City and will fill a definite public need and will be of great value to the City; and

WHEREAS, the City is also in need of a suitable place for the holding of various events, including but not limited to amateur baseball games and similar activities; and

WHEREAS, Ball Club is able to have the baseball stadium and grounds known and Wrigley Field conveyed to City; and

WHEREAS, said Wrigley Field would be suitable for such purposes and would be of great value to the City and its inhabitants; and

WHEREAS, Ball Club is willing to cause Wrigley Field to be conveyed to the City for said property now owned and to be acquired as aforesaid by the City in Chavez Ravine; and

WHEREAS, the bringing of said Major League Baseball Club to Los Angeles would result in a large additional payroll in this area; and

WHEREAS, all of the foregoing is useful and convenient in connection with the exercise of the City's rights and powers and is in the public interest,

NOW, THEREFORE, IT IS UNDERSTOOD AND AGREED AS FOLLOWS, TO WIT:

1. City will convey the property presently owned by it in Chavez Ravine consisting of 185 acres, more or less, and will use its best efforts to acquire, at a reasonable cost, and convey additional land, to make a total of 300 acres, more or less, all as shown in Exhibit A, to Ball Club, or nominee; provided, however:

(a) City shall reserve all mineral rights which it now owns or may hereafter acquire, and a suitable drill site for the production of oil from said property, the location of such site to be mutually agreed upon by the parties hereto, which location shall not interfere with Ball Club's operations, and not exceed 5 acres in size.

(b) One-half of all monies, payments, royalties, or other consideration received by City from said mineral rights, or any of them, in whole or in part, shall be placed in a special trust fund by City, and such funds shall be expended solely for the purpose of providing and maintaining recreational facilities to promote the youth program of Ball Club, the location and type of such recreational facilities to be mutually agreed upon by the parties hereto, with approval by the City Council by ordinance.

(c) That title to said 40 acres thereof, the location of such 40 acres to be designated by Ball Club, shall be retained by the City for a period of 20 years to assure performance by Ball Club of its policy of providing and maintaining recreational facilities. In the event such policy, and all of the terms of this agreement pertaining to the recreational facilities on said 40 acres, shall have been fully and faithfully performed for a period of 20 years, title of such 40 acres shall be conveyed to the Ball Club, or nominee, forthwith, without further consideration.

It is understood and agreed that any violation of the terms of the agreement with reference to the recreational facilities shall not invalidate or affect any transfers of land which may theretofore have been made pursuant to this agreement.

(d) All property herein agreed to be conveyed by City, including said 40 acres, will be conveyed free and clear of any deed restriction on use, and title policy shall be furnished to Ball Club. In this connection, it is expressly understood and agreed that the portion of property described in paragraph 1 now owned by the City and which was acquired by City from the Los Angeles Housing Authority has a deed restriction reading, in part, as follows:

"To be used for public purpose only; and not to be used directly or indirectly by the City of Los Angeles, or its grantees, successors in interest, assigns, or any other person or persons whatsoever claiming by, through or under the City of Los Angeles, for a period of 20 years from and after the date hereof, for residential development or residential subdivision."

City agrees to use its best efforts to have such deed restrictions eliminated or modified so as to permit the use contemplated by Ball Club; provided that in the event City is unable to have such restriction so eliminated or modified, this contract shall be of no further force or effect.

2. Upon conveyance of such property to Ball Club, the existing public streets therein which would no longer be needed for present or future street purposes will be vacated and the City shall, upon demand of Ball Club, commence proceedings to vacate said streets and deliver any title which may remain in the City without further consideration.

3. Prior to passage of title to any of the acreage described in paragraph 1, City shall spend up to but not to exceed $2,000,000 to place such property in a proper condition to convey to Ball Club or its nominee, the manner in which such money will be spent for such purpose to be designated by Ball Club.

4. City agrees to provide such public streets as may be needed within the periphery of the area to be acquired by Ball Club; provided, however, that the cost thereof, other than the cost of acquiring necessary land, shall be considered a part of the $2,000,000 which the City is to spend to prepare the site for sale.

5. Ball Club shall cause to be conveyed to City the land and improvements now known as Wrigley Field, including all mineral rights; provided, however, that one-half of all monies, payments, royalties, or other consideration received by City from said mineral rights, or any of them, in

whole or in part, shall be placed in a special trust fund by City and such funds shall be expanded solely for the purpose of providing and maintaining recreational facilities to promote the youth program of Ball Club, the location and type of such recreational facilities to be mutually agreed upon by the parties hereto, with approval by the City Council by ordinance; further provided, that Ball Club reserves the right to the use of said Wrigley Field until the stadium referred to in paragraph 6 shall be completed and ready for use, conditioned upon payment to City of a rental to be mutually agreed upon.

6. Ball Club shall cause to be constructed on property conveyed by City, at Ball Club's cost and expense, a modern baseball stadium, seating not less than 50,000 people.

7. Ball Club will cause to be moved to the City of Los Angeles the present Brooklyn National League Baseball franchise and ball team known as the "DODGERS".

The City of Los Angeles agrees that it will initiate proceedings for the purpose of rezoning said property to "C-3" and for the granting of a conditional use permitting its use for a baseball stadium as herein provided.

It is recognized that the method or means of carrying out the terms and conditions of this contract in detail have not been provided herein in every instance but that the parties hereto will use their best efforts in arriving at mutually acceptable methods of procedure and modes of operation as to such details. Any such action on the part of the City shall be submitted to the City Council for approval.

IT IS EXPRESSLY UNDERSTOOD AND AGREED that this agreement is made in reliance upon the action taken by the Board of Supervisors of the County of Los Angeles in Resolution of said Board of Supervisors adopted and entered in the minutes of said Board on Tuesday, September 17, 1957, a copy of which said resolution is attached hereto, marked Exhibit "B", and made a part hereof, providing for the furnishing of funds to construct necessary access roads, including cost of acquiring any required rights of way, and the City agrees that upon the County's making such funds available to City it will diligently proceed with the construction of such access roads. City agrees that it will make demand as required upon the County of Los Angeles to furnish the funds heretofore voted and/or appropriated for said access roads.

IN WITNESS WHEREOF, The City of Los Angeles has caused this instrument to be executed in its behalf by its duly authorized officers, and the Ball Club has executed the same

by its duly authorized officers and has caused its corporate seal to be hereunto affixed, all on the day and year
first hereinabove written.

THE CITY OF LOS ANGELES

By _____
Acting Mayor

(SEAL)

ATTEST:

_____
City Clerk

LOS ANGELES DODGERS, INC. (formerly
known as the BROOKLYN NATIONAL
LEAGUE BASEBALL CLUB, INC.)

By _____
President

(SEAL)

By _____
Secretary

E X H I B I T " B "

Gordon T. Nesvig, Chief Clerk of the Board.
COUNTY OF LOS ANGELES BOARD OF SUPERVISORS, 501 HALL OF
RECORDS, LOS ANGELES 12.
Members of the Board: John Anson Ford, Chairman; Herbert
C. Legg, Kenneth Hahn, Burton W. Chace and Warren M. Dorn.

R E S O L U T I O N

Introduced by Supervisors
Tuesday, September 17, 1957
The Board met in regular session. Present: Supervisors
John Anson Ford, Chairman presiding, Herbert C. Legg, Ken-
neth Hahn, Burton W. Chace and Warren M. Dorn; and Harold
J. Ostly, Clerk, by Gordon T. Nesvig, Deputy Clerk.

* * * * *

IN RE MAJOR LEAGUE BASEBALL IN COUNTY OF LOS ANGELES:
RESOLUTION DETERMINING THAT COUNTY OF LOS ANGELES WILL MAKE

AVAILABLE 2,740,000.00 TO CITY OF LOS ANGELES FOR PUBLIC
APPROACH ROAD IMPROVEMENTS TO THE CHAVEZ RAVINE AREA AND
INSTRUCTING CHIEF ADMINISTRATIVE OFFICER AND ROAD COMMIS-
SIONER RELATING TO FUNDS REQUIRED.

On motion of Supervisor Hahn, duly carried by the following vote, to wit: Ayes: Supervisors Legg, Hahn, Chace,
Dorn and Ford; Noes, none, it is ordered that the following resolution be and the same is hereby adopted:

WHEREAS, the Board of Supervisors of the County of Los
Angeles believes that Major League Baseball would be a
recreational and economic asset to this community; and

WHEREAS, both the City and the County have entered into
negotiations with the Brooklyn Dodgers, looking toward the
transfer of that Club's franchise from New York to Los Angeles; and

WHEREAS, these negotiations contemplate that a site in the
Chavez Ravine area will be made available for the construction by the Brooklyn Ball Club of a modern baseball
stadium – with parking facilities for 25,000 cars; and

WHEREAS, the Brooklyn Ball Club is also proposing to develop on the site a regional sports center covering approximately 40 acres, these recreational facilities to be
available for free public use; and

WHEREAS, if this total development is accomplished, there
will be a major increase in vehicular traffic in the Chavez
Ravine area because of the public use of the proposed facilities; and

WHEREAS, traffic engineering studies show that present
access streets to the Chavez Ravine site are entirely inadequate to accommodate the public use thereof, as contemplated:

NOW, THEREFORE, BE IT RESOLVED that the County of Los
Angeles determines that if the proposed improvements are
located in the Chavez Ravine area as contemplated the public necessity and convenience will require approach road
improvements to the area, and that such improvements will
be of general county interest; and

BE IT FURTHER RESOLVED that if a site in the Chavez Ravine area is made available by the City of Los Angeles for
these purposes – and concrete evidence of its contemplated
or actual improvement in the manner and for the purposes
referred to herein is presented to the County, the County
of Los Angeles will make available in the manner provided
by law a sum of money not to exceed $2,740,000.00 or as much
thereof as is needed, to the City of Los Angeles payable on
demand by the City, for public approach road improvements
to the Chavez Ravine area; and

BE IT FURTHER RESOLVED that the Board of Supervisors
hereby instruct the Chief Administrative Officer and the

Road Commissioner to include within the Road Department 1958–59 budget (Motor Vehicle Fund) for the purposes recited herein an appropriation in an amount estimated to be required for expenditure in 1958–59, and in subsequent years as required, appropriations representing the balance required out of the total commitment of $2,740,000.00, provided that such balance be further reduced by any funds which may be advanced during the current fiscal year by action of the Board of Supervisors.

I hereby certify that the foregoing is a full, true and correct copy of a resolution which was adopted by the Board of Supervisors of the County of Los Angeles, State of California, on September 17, 1957, and entered in the minutes of said Board.

> HAROLD J. OSTLY, County Clerk of the County of Los Angeles, State of California, and ex officio Clerk of the Board of Supervisors of said County.
>
> By /s/ VIRGINIA MULLENDORE
> Deputy Clerk

(SEAL)

The Dodgers' Record in Los Angeles

Year	Record	Finish	Margin	Attendance
1958	71–83	7th	−21	1,845,556
1959	88–68	1st	+2	2,071,045
1960	82–72	4th	−13	2,253,887
1961	89–65	2nd	−4	1,804,250
1962	102–63	2nd	−1	2,755,184
1963	99–63	1st	+6	2,538,602
1964	80–82	6th	−13	2,228,751
1965	97–65	1st	+2	2,553,577
1966	95–67	1st	+1½	2,617,029
1967	73–89	8th	−28½	1,664,362
1968	76–86	7th	−21	1,581,093
1969*	85–77	4th	−8	1,784,527
1970	87–74	2nd	−14½	1,697,142
1971	89–73	2nd	−1	2,064,594
1972	85–70	3rd	−10½	1,860,858
1973	95–66	2nd	−3½	2,136,192
1974	102–60	1st	+4	2,632,474
1975	88–74	2nd	−20	2,539,349
1976	92–70	2nd	−10	2,386,301
1977	98–64	1st	+10	2,955,087
1978	95–67	1st	+2½	3,347,845
1979	79–83	3rd	−11½	2,860,954
1980	92–71	2nd	−1	3,249,287
1981**	63–47	1st		2,381,292
1982	88–74	2nd	−1	3,608,881
1983	91–71	1st	+2	3,510,313
1984	79–83	4th	−13	3,134,824
1985	95–67	1st	+5½	3,264,593
1986	73–89	5th	−23	3,023,208

*Division play begins
**Strike shortened season

Notes

Chapter 1

1. Damon Rice, *Seasons Past* (New York: Praeger, 1976), p. 432.
2. *New York Daily News*, Sept. 30, 1957.
3. *Brooklyn Eagle*, Apr. 5, 1913.
4. Ibid., Mar. 6, 1912.
5. Ibid.
6. David Quentin Voigt, *American Baseball*, 3 vols. (Penn State Univ. Press, 1983), vol. 1, chap. 1.
7. Ibid., vol. 1, chap. 8.
8. Frank Graham, *The Brooklyn Dodgers* (New York: Putnam, 1945), pp. 10-13.
9. These cities include Altoona, Buffalo, Columbus, Hartford, Indianapolis, Louisville, Providence, Richmond, Rochester, N.Y., St. Paul, Syracuse, Toledo, Troy, N.Y., Wilmington, and Worcester.
10. Voigt, *American Baseball*, vol. II, chap. 5.
11. Ibid., p. 143.
12. *Brooklyn Eagle*, Oct. 8, 1916.
13. Edward Robb Ellis, *The Epic of New York City* (New York: Coward-McCann, 1966), chap. 42.
14. *Brooklyn Eagle*, Sept. 14, 1920.
15. Graham, *Brooklyn Dodgers*, p. 131.
16. Rice, *Seasons Past*, p. 304.
17. Graham, *Brooklyn Dodgers*, pp. 150-51.
18. Ibid., p. 157.
19. Roger Kahn, *The Boys of Summer* (New York: Harper and Row, 1971).
20. See Harold Rosenthal, *The Ten Best Years of Baseball* (Chicago: Contemporary Books, 1979) and Harvey Frommer, *New York City Baseball: The Last Golden Age: 1947–57* (New York: MacMillan, 1980).
21. Ira Berkow, *Red: A Biography of Red Smith—The Life and Times of a Great American Writer* (New York: Times Books, 1986), p. 92.
22. *Brooklyn Eagle*, Oct. 6, 1941.
23. Ibid.

24. Ibid., Oct. 4, 1951.
25. Ibid.
26. Peter Golenbock, *Bums* (New York: Putnam, 1984), p. 406.
27. Ibid., pp. 407–8.
28. Robert Wagner Papers, New York City Municipal Archives.

Chapter 2

1. Frommer, *New York City Baseball*, p. 219.
 2. The principal source of information on Branch Rickey is Murray Polner, *Branch Rickey* (New York: Signet, 1982).
 3. For Rickey's playing career see Joseph L. Reichler, ed., *The Baseball Encyclopedia*, 6th ed. (New York: Macmillan, 1985), p. 1328.
 4. Polner, *Branch Rickey*, p. 79.
 5. Reichler, *Baseball Encyclopedia*, pp. 269–90.
 6. Red Smith, *To Absent Friends* (New York: Atheneum, 1983), p. 447.
 7. Voigt, *American Baseball*, vol. II, p. 87.
 8. Smith, *Absent Friends*, p. 447.
 9. Voigt, *American Baseball*, vol. II, pp. 138–48.
10. Polner, *Branch Rickey*, p. 113.
11. Maury Allen, *Baseball's 100* (New York: Galahad Books, 1981), pp. 33–34.
12. Polner, *Branch Rickey*, p. 146.
13. Ibid.
14. See especially Jules Tygiel, *Baseball's Great Experiment: Jackie Robinson and His Legacy* (New York: Oxford Univ. Press, 1983), and Harvey Frommer, *Rickey and Robinson* (New York: Macmillan, 1982).
15. Tygiel, *Great Experiment*, pp. 30–32.
16. Ibid., p. 81.
17. Voigt, *American Baseball*, vol. III, pp. 111–12, and Polner, *Branch Rickey*, pp. 253–56.
18. Polner, *Branch Rickey*, pp. 216–17.
19. Ibid., p. 249.
20. Ibid.
21. Smith, *Absent Friends*, p. 377.
22. Kahn, *Boys of Summer*, p. 424.
23. *Sporting News*, Oct. 28, 1955.
24. Kahn, *Boys of Summer*, p. 427.
25. Polner, *Branch Rickey*, p. 153.
26. Ibid., p. 215.
27. Ibid., pp. 220–21.
28. Smith, *Absent Friends*, p. 376.
29. Polner, *Branch Rickey*, p. 216.
30. Kahn, *Boys of Summer*, p. 426.
31. Ibid., p. 428.
32. *Brooklyn Eagle*, Sept. 29, 1916.
33. Lowell Reidenbaugh, *Take Me Out To The Ball Park* (St. Louis: Sporting News, 1983), p. 393.
34. Kahn, *Boys of Summer*, p. xiv.
35. Ibid., p. 35.

36. Voigt, *American Baseball*, vol. III, pp. 103–4.
37. Ibid., pp. 314–21.
38. Voigt, vol. III, *American Baseball*, chap. 5.
39. *Organized Professional Team Sports*. Hearings Before the Antitrust Subcommittee of the Committee on the Judiciary, House of Representatives, 85th Cong. 1st sess., p. 2047. Hereafter: Antitrust 1957.
40. *Sporting News*, Sept. 14, 1955.
41. Ibid.
42. Antitrust 1957, p. 2046.
43. *Sporting News*, Aug. 31, 1955.
44. Antitrust 1957, p. 1868.
45. Ibid., pp. 1874–75.
46. Ibid., p. 1881.
47. *U.S. News and World Report*, Oct. 15, 1984.
48. *New York Times*, Mar. 6, 1952.
49. Frommer, *New York City Baseball*, p. 2.
50. Ibid., p. 3.
51. Ibid.
52. Ibid.
53. Golenbock, *Bums*, pp. 432–33.
54. Ibid., p. 433.
55. Rice, *Seasons Past*, p. 405.
56. Antitrust 1957, p. 2053.
57. *New York Times*, Mar. 4, 1953.
58. Antitrust 1957, p. 2053.
59. Kahn, *Boys of Summer*, p. 428.

Chapter 3

1. David McCullough, *The Great Bridge* (New York: Simon and Schuster, 1972), pp. 450–455.
2. Ellis, *Epic of New York*, pp. 450–455.
3. Wallace Sayre and Herbert Kaufman, *Governing New York City: Politics in the Metropolis* (New York: Norton, 1965), p. 11.
4. Ibid., p. 12.
5. Ellis, *Epic of New York*, p. 452.
6. Sayre and Kaufman, *Governing New York*, pp. 327–57.
7. Ellis, *Epic of New York*, p. 452.
8. Sayre and Kaufman, *Governing New York*, p. 16.
9. James Madison, *Federalist Fifty-One* (New York: New American Library, 1964), p. 322.
10. Ellis, *Governing New York*, pp. 579–92.
11. See Robert Caro, *The Power Broker* (New York: Random House, 1975).
12. The Housing Act of 1949—P.L. 171, the Federal Housing Act of 1949.
13. Letter from Robert Moses to Walter O'Malley, Aug. 15, 1955, Wagner Papers.
14. Letter from Robert Moses to John Cashmore, Aug. 26, 1955, Wagner Papers.
15. *New York Times*, Aug. 20, 1955.

16. Caro, *Power Broker*, p. 777.
17. *New York Times*, Aug. 18, 1955.
18. Letter from Robert Moses to John Cashmore, Aug. 16, 1955, Wagner Papers.
19. Ira Rosenwaike, *Population History of New York City* (Syracuse Univ. Press, 1972), p. 137.
20. Golenbock, *Bums*, p. 428.
21. Kahn, *Boys of Summer*, p. xv.
22. *New York Times*, Aug. 18, 1955.
23. Ibid., p. 26.
24. New York City Board of Estimate, Journal of Proceedings, July 22–Sept. 22, 1955, vol. XII, pp. 10,881–82.
25. *New York Times*, Aug. 20, 1955.
26. Board of Estimate, p. 10,881.
27. Ibid., p. 10,883.
28. Ibid.
29. Letter from Robert Moses to Walter O'Malley, Aug. 15, 1955, Wagner Papers.
30. Golenbock, *Bums*, p. 429.
31. Ibid., p. 19.
32. Antitrust 1957, p. 1855.
33. Ibid., 1854.

Chapter 4

 1. Kahn, *Boys of Summer*, pp. 170–79.
 2. Rice, *Seasons Past*, p. 394.
 3. *New York Times*, Apr. 14, 1955. All subsequent accounts of the opening weeks of the 1955 season are based on reports in the *New York Times*.
 4. Ibid., Apr. 22, 1955.
 5. Ibid., May 7, 1955.
 6. Reichler, *Baseball Encyclopedia*, p. 415.
 7. Golenbock, *Bums*, pp. 403–4.
 8. *New York Times*, Oct. 1, 1955.
 9. Antitrust 1957, p. 2053.
10. Ibid.
11. Frommer, *New York City Baseball*, p. 219.
12. Ibid.
13. Antitrust 1957, p. 2053.
14. Ibid.
15. Ibid.
16. *Sporting News*, Aug. 17, 1955.
17. Ibid, Aug. 24, 1955.
18. Ibid, Dec. 14, 1955.
19. Ibid.
20. *Sporting News*, Dec. 14, 1955.
21. Ibid.
22. Ibid.
23. Ibid.

24. *New York Times*, Nov. 4, 1953.
25. Ibid., Feb. 7, 1956.
26. Ibid., Feb. 8, 1956.
27. Ibid., Feb. 9, 1956.
28. Ibid., Mar. 2, 1956.
29. Ibid., Mar. 9, 1956.
30. Ibid., Mar. 30, 1956.
31. Ibid., Apr. 7, 1956.
32. Ibid., Apr. 9, 1956.
33. Ibid., May 24, 1956.
34. *A Planning Study For the Area Bounded by Vanderbilt Avenue, De Kalb Avenue, Sterling Place, and Bond Street in the Borough of Brooklyn, New York.* Nov. 20, 1956, Gilmore D. Clarke and Michael Rapuano.
35. *New York Times*, July 23, 1956.
36. Ibid., July 25, 1956.
37. Ibid.
38. Ibid., Oct. 1, 1956.
39. Ibid., Sept. 27, 1956.
40. Ibid., Oct. 31, 1956.
41. Antitrust 1957, p. 1873.
42. Frommer, *New York City Baseball*, p. 219.
43. Antitrust 1957, p. 2046.
44. *New York Times*, Nov. 29, 1956. See note 34 *supra*.
45. Memo from Robert Moses to Robert Wagner, Dec. 7, 1956, Wagner Papers.
46. Ibid.
47. Ibid.
48. *New York Times*, Dec. 24, 1956.
49. Letter to John Cashmore, Clarke-Rapuano Report, Nov. 20, 1956.

Chapter 5

1. John D. Weaver, *Los Angeles: The Enormous Village, 1781–1981* (Santa Barbara: Capra Press, 1980), chap. 1.
2. Ibid., p. 29.
3. See Kevin Starr, *Inventing the Dream: California Through The Progressive Era* (New York: Oxford Univ. Press, 1985), chap. 7; Matthew Josephson, *The Robber Barons* (New York: Harcourt Brace Jovanovich), 1962, chap. 4.
4. Robert M. Fogelson, *The Fragmented Metropolis: Los Angeles, 1850–1930* (Cambridge: Harvard Univ. Press, 1967), chap. 10, and Starr, *The Dream*, chap. 8.
5. Fogelson, *Fragmented Metropolis*, p. 211.
6. Such fusion tickets were successfully forged by Fiorello LaGuardia in 1933 and by John Lindsay in 1969.
7. John C. Ries and John J. Kirlin, "Government in the Los Angeles Area: The Issue of Centralization and Decentralization," in Werner A. Hirsch, *Los Angeles: Viability and Prospects for Metropolitan Leadership* (New York: Praeger, 1971), p. 92.
8. Ibid., pp. 93–95.

9. Norris Poulson, *Memoirs*, Department of Special Collections, Research Library, Univ. of California at Los Angeles, p. 199.
10. Ibid.
11. Thomas S. Hines, "Housing, Baseball, and Creeping Socialism: The Battle of Chavez Ravine, Los Angeles 1949–1959," *Journal of Urban History*, Feb. 1982, p. 124.
12. Ibid., p. 137.
13. Ibid., p. 139.
14. Ibid., p. 138.
15. Ibid., p. 139.
16. *Housing Authority v. City of Los Angeles*, 38 C.2d 853, Apr. 1952, p. 866.
17. Ibid., p. 867.
18. Poulson, *Memoirs*, p. 199.
19. Author interview with Roger Arneburgh, Jan. 22, 1986.
20. Poulson, *Memoirs*, p. 199.
21. Arneburgh interview.
22. Ibid.
23. Author interview with Kenneth Hahn, Jan. 22, 1986.
24. Cary S. Henderson, "Los Angeles and the Dodger War, 1957–62," *Southern California Quarterly*, Fall 1980, p. 264. This excellent article is one of the few analyses of the Los Angeles side of the Dodgers' move.
25. Poulson, *Memoirs*, p. 197.
26. Ibid.
27. Ibid.
28. Fred W. Lange, *History of Baseball in California and Pacific Coast Leagues 1847–1938* (Oakland, n.p., 1938), p. 14.
29. *Sporting News*, Aug. 31, 1949.
30. See Richard Goldstein, *Spartan Seasons: How Baseball Survived the Second World War* (New York: Macmillan), 1980.
31. Voigt, *American Baseball*, vol. III, pp. 206–7.
32. *New York Times*, Dec. 5, 1945.
33. *Sporting News*, Dec. 20, 1945.
34. *New York Times*, Dec. 8, 1946; *Sporting News*, Dec. 11, 1946.
35. *New York Times*, Aug. 12, 1947.
36. U.S. House of Representatives, *Organized Baseball*, Report of the Subcommittee on the Study of Monopoly Power of the Committee of the Judiciary Pursuant to H. Res. 95, 82 Cong. 2nd sess., pp. 142–46.
37. *Sporting News*, Aug. 13, 1947.
38. Ibid.
39. *New York Times*, Nov. 13, 1947.
40. Ibid., Dec. 5, 1947.
41. Ibid., Dec. 12, 1947.
42. Ibid., Apr. 4, 1950.
43. Ibid., Nov. 2, 1950.
44. Ibid., Aug. 30, 1951; *Sporting News*, Sept. 5, 1951, and Sept. 12, 1951.
45. *New York Times*, Aug. 31, 1951.
46. *Sporting News*, Dec. 5, 1951.
47. Ibid.
48. Arneburgh interview. See also *Los Angeles Times*, Aug. 25, 1963.

49. *Los Angeles Times*, Feb. 22, 1957.

50. Antitrust 1957, p. 1,861.

51. Golenbock, *Bums*, p. 442 and Harold Parrott, *The Lords of Baseball* (New York: Praeger, 1976), p. 11.

52. Antitrust 1957, p. 1887.

53. *Los Angeles Times*, Feb. 2, 1957.

54. Poulson, *Memoirs*, p. 200.

55. Ibid.

56. Ibid.

57. Ibid.

58. *Los Angeles Times*, May 2, 1957.

59. Poulson, *Memoirs*, p. 201.

60. Ibid.

61. Ibid.

62. Author interview with Dick Walsh, Jan. 16, 1986.

63. Hahn interview.

64. Author interview with Red Patterson, Jan. 23, 1986.

65. Poulson, *Memoirs*, p. 201.

66. Ibid., pp. 201–2.

67. *Los Angeles Times*, Aug. 27, 1957.

68. On Aug. 19, 1957, the Giants announced they would move to San Francisco for the 1958 season.

69. *Los Angeles Times*, Aug. 28, 1957.

70. Ibid., Sept. 12, 1957.

71. Poulson, *Memoirs*, p. 203.

72. *Los Angeles Times*, June 13, 1957.

73. Poulson, *Memoirs*, p. 204.

74. *Los Angeles Times*, Sept. 17, 1957.

75. Ibid.

76. Ibid.

77. Ibid.

78. Ibid.

79. Ibid., Sept. 29, 1957.

80. Ibid.

81. Ibid., Sept. 30, 1957.

82. Ibid., Oct. 1, 1957.

83. Poulson, *Memoirs*, p. 203.

84. Ibid.

85. *Los Angeles Times*, Oct. 1, 1957.

86. Ibid.

87. Poulson, *Memoirs*, p. 204.

88. *Los Angeles Times*, Oct. 2, 1957.

89. Ibid., Oct. 7, 1957.

90. Author interview with Rosalind Wyman, Jan. 17, 1986.

91. *Los Angeles Times*, Oct. 8, 1957.

92. Wyman interview.

Chapter 6

1. *New York Times*, Feb. 22, 1957.
2. Ibid.
3. Ibid.
4. *Sporting News*, Jan. 9, 1957.
5. *Los Angeles Times*, Feb. 22, 1957.
6. Ibid., Mar. 8, 1957.
7. Ibid., Mar. 12, 1957.
8. *New York Times*, Mar. 25, 1957.
9. Ibid., Mar. 26, 1957.
10. Ibid.
11. Ibid., Apr. 19, 1957.
12. Ibid.
13. Caro, *Power Broker*, p. 1083.
14. *New York Times*, Apr. 20, 1957.
15. Ibid.
16. Memo from John Larkin to John Theobald, Jul. 19, 1957, Wagner Papers.
17. *New York Times*, May 17, 1957.
18. Ibid.
19. Ibid., May 23, 1957.
20. Reichler, *Baseball Encyclopedia*, pp. 424–25.
21. *New York Times*, May 29, 1957.
22. Ibid.
23. Ibid., June 1, 1957.
24. Ibid., May 30, 1957.
25. Golenbock, *Bums*, p. 439.
26. Rice, *Seasons Past*, p. 427.
27. Golenbock, *Bums*, p. 439.
28. Frommer, *New York City Baseball*, p. 7.
29. Rice, *Seasons Past*, p. 418.
30. Voigt, *American Baseball*, vol. III, p. 87.
31. Frommer, *New York City Baseball*, p. 219.
32. Antitrust 1957, p. 2053.
33. Ibid., p. 2047.
34. Ibid., p. 2046.
35. Frommer, *New York City Baseball*, p. 8.
36. Golenbock, *Bums*, p. 442.
37. Author interview with Chub Feeney, July 24, 1957.
38. Letter from George McLaughlin to Horace Stoneham, June 6, 1957, Wagner Papers.
39. McLaughlin Baseball Plan, p. 1., Wagner Papers.
40. *New York Times*, June 21, 1957.
41. Letter from Horace Stoneham to George McLaughlin, June 6, 1957, Wagner Papers.
42. Letter from George McLaughlin to Horace Stoneham, June 11, 1957, Wagner Papers.
43. Memo, June 13, 1957, Wagner Papers.
44. *New York Times*, June 21, 1957.

45. Telegram from William Shea to Warren Giles, June 20, 1957, Wagner Papers.
46. Letter from George McLaughlin to Warren Giles, June 24, 1957, Wagner Papers.
47. Ibid.
48. Ibid.
49. *Los Angeles Times,* June 12, 1957.
50. Ibid.
51. *Federal Baseball Club of Baltimore, Inc. v. National League of Professional Baseball Clubs, et. al.,* 259 U.S. 200 (1922).
52. *Federal Baseball,* pp. 208–9.
53. *American League Baseball Club v. Chase* (1914).
54. James B. Dworkin, *Owners Versus Players: Baseball and Collective Bargaining* (Boston: Auburn House, 1981), pp. 59–62.
55. *George Toolson v. New York Yankees, Inc.,* 346 U.S. 356 (1953).
56. *Toolson,* p. 357.
57. Ibid.
58. *Radovich v. National Football League* (1957), p. 452.
59. *Radovich,* p. 452.
60. Ibid.
61. *Curtis R. Flood v. Bowie Kuhn,* 407 U.S. 258 (1972), p. 284.
62. Antitrust 1957, p. 1359.
63. Ibid., p. 1365.
64. Ibid.
65. Ibid.
66. Ibid., p. 1366.
67. Ibid., p. 1850. Walter O'Malley's testimony can be found on pp. 1850–86.
68. Ibid., p. 1853.
69. Ibid., p. 1854.
70. Ibid., p. 1856.
71. Ibid. The Montreal reference is to the Dodgers' Triple A farm team.
72. Ibid., p. 1857.
73. Ibid., pp. 1857–58.
74. Ibid., p. 1858.
75. Ibid., p. 1858.
76. Ibid.
77. Ibid., p. 1859.
78. Ibid., p. 1860.
79. Ibid.
80. Ibid.
81. Ibid., pp. 1860–61.
82. Ibid., p. 1861.
83. Ibid., p. 1862.
84. Ibid., p. 1864.
85. Ibid., p. 1865.
86. Ibid., p. 1872.
87. Ibid., pp. 1872–73.
88. *New York Times,* July 18, 1957.
89. Ibid., July 19, 1957.

90. Ibid.
91. "Robert Moses on the Battle of Brooklyn," *Sports Illustrated,* July 11, 1957, pp. 26–49.
92. Ibid., p. 27.
93. Ibid.
94. Ibid., p. 48.
95. *New York Times,* July 18, 1957.
96. Ibid.
97. Madigan-Hyland Report, "Brooklyn Stadium and Related Developments," Aug. 5, 1957.
98. Ibid., p.i.
99. *New York Times,* Aug. 7, 1957.
100. Letter from Robert Moses to John Theobald, Aug. 19, 1957, Wagner Papers.
101. *New York Times,* Aug. 27, 1957.
102. Memo from Charles Riedel to John Cashmore, Aug. 29, 1957, Wagner Papers.
103. Ibid.
104. *New York Times,* Aug. 27, 1957.
105. Letter from Robert Moses to Peter Brown, Aug. 26, 1957, Wagner Papers.
106. *New York Times,* Aug. 20, 1957.
107. Letter from Jay Coogan to Robert Moses, Aug. 29, 1957, Wagner Papers.
108. *New York Times,* Aug. 28, 1957.
109. Ibid.
110. Ibid., Sept. 6, 1957.
111. Ibid., Sept. 9, 1957.
112. Ibid., Sept. 11, 1957.
113. Ibid.
114. Ibid.
115. Ibid.
116. Ibid., Sept. 19, 1957.
117. Letter from Robert Moses to William Peer, Sept. 20, 1957, Wagner Papers.
118. Transcript, Robert Moses to Peter Brown, Sept. 10, 1957, Wagner Papers.
119. *New York Times,* Sept. 21, 1957.
120. Ibid.
121. Letter from Robert Moses to Robert Wagner, Sept. 25, 1957, Wagner Papers.
122. *New York Times,* Oct. 9, 1957.

Chapter 7

1. *New York Times,* Oct. 14, 1957.
2. Ibid.
3. See Golenbock, *Bums;* Kahn, *Boys of Summer;* Rice, *Seasons Past.*
4. *New York Times,* Nov. 23, 1957.
5. Henderson, pp. 275–76.
6. *Los Angeles Times,* Aug. 10, 1979.
7. *New York Times,* Dec. 7, 1956.
8. Ibid.

9. *Los Angeles Times,* Jan. 5, 1958.
10. Ibid., Jan. 3, 1958.
11. Ibid., Jan. 8, 1958.
12. Ibid., Jan. 12, 1958.
13. Ibid.
14. Ibid., Jan. 14, 1958.
15. Ibid.
16. Ibid., Jan. 15, 1958.
17. Ibid., Jan. 17, 1958.
18. Ibid., Jan. 18, 1958.
19. Hahn interview.
20. Parrott, *Lords of Baseball,* p. 241.
21. Reidenbaugh, *Take Me Out,* pp. 142–43.
22. Parrott, *Lords of Baseball,* p. 243.
23. *Los Angeles Times,* Jan. 20, 1958.
24. Ibid., Oct. 17, 1957.
25. See Erwin G. Krasnow, Lawrence D. Longley, Herbert A. Terry, *The Politics of Broadcast Regulation,* 3rd ed. (New York: St. Martin's Press, 1982).
26. Poulson, *Memoirs,* p. 204.
27. Antitrust 1957, pp. 2110–27.
28. *Los Angeles Times,* July 31, 1957.
29. *New York Times,* Feb. 23, 1958.
30. Ibid., Apr. 16, 1958.
31. *Los Angeles Times,* Apr. 28, 1958.
32. Ibid.
33. Poulson, *Memoirs,* p. 205.
34. Author interview with Vin Scully, Mar. 26, 1986.
35. *Los Angeles Times,* Mar. 20, 1958.
36. Ibid., Mar. 15, 1958.
37. Ibid., June 3, 1958.
38. Statistics taken from Reichler, *Baseball Encyclopedia.*
39. In 1955–57, Bilko had a total of 148 homers and 428 r.b.i.'s for the Los Angeles Angels of the Pacific Coast League.
40. *Los Angeles Times,* Apr. 16, 1958.
41. *New York Times,* Apr. 26, 1958.
42. Gene Schoor, *The Complete Dodgers Record Book* (New York: Facts on File, 1984), p. 418.
43. *Los Angeles Times,* Oct. 28, 1958.
44. Ibid., Sept. 28, 1958.
45. Ibid.
46. Ibid., Sept. 29, 1958.
47. Ibid., Sept. 2, 1958. This prediction was fulfilled in 1978 when the Dodgers drew 3,347,845.
48. Poulson, *Memoirs,* p. 204.
49. Ibid.
50. Ibid., p. 205.
51. Ibid.
52. *Los Angeles Times,* Apr. 15, 1958.
53. Ibid., May 16, 1958.

54. Wyman interview.
55. *Hearings on Chavez Ravine,* Assembly Interim Committee on Governmental Efficiency and Economy, May 15, 1958, pp. 39–40. Hereafter: Assembly Hearings.
56. Ibid., pp. 40–41.
57. Ibid., p. 41.
58. *Los Angeles Times,* May 17, 1958.
59. Ibid.
60. Ibid.
61. Ibid.
62. Ibid., May 18, 1958.
63. Ibid., May 23, 1958.
64. Ibid.
65. Ibid.
66. Ibid., May 24, 1958.
67. Ibid.
68. Poulson, *Memoirs,* p. 205.
69. Ibid.
70. *Los Angeles Times,* May 27, 1958.
71. Ibid.
72. Ibid., May 28, 1958.
73. Ibid.
74. Antitrust 1957, p. 1852.
75. *Los Angeles Times,* May 28, 1958.
76. Ibid., May 29, 1958.
77. Ibid.
78. Ibid.
79. Ibid.
80. Ibid., June 1, 1958.
81. Ibid.
82. Ibid., June 2, 1958.
83. Ibid.
84. Ibid., June 3, 1958.
85. Ibid., June 5, 1958.
86. Ibid.
87. Ibid.
88. Ibid.
89. Ibid.

Chapter 8

1. *Reuben v. City of Los Angeles,* Los Angeles Superior Court No. 687210 (1958), and *Kirshbaum v. Housing Authority,* Los Angeles Superior Court No. 699077 (1958).
2. *Los Angeles Times,* June 5, 1958.
3. Ibid., June 11, 1958.
4. Ibid., June 21, 1958.
5. Phill Silver, "Points and Authorities in Support of Plaintiff's Application for a Temporary Restraining Order." Los Angeles Superior Court (1958).

6. *Los Angeles Times,* June 21, 1958.
7. Ibid.
8. Ibid.
9. Ibid.
10. Ibid.
11. Ibid.
12. Golenbock, *Bums,* p. 441.
13. *Los Angeles Times,* June 21, 1958.
14. Ibid., June 24, 1958.
15. Ibid.
16. Ibid., June 25, 1958.
17. Ibid.
18. Ibid., June 26, 1958.
19. Ibid.
20. Ibid., July 7, 1958.
21. Ibid.
22. Ibid.
23. Ibid.
24. Arneburgh interview.
25. *Los Angeles Times,* July 15, 1958.
26. Ibid.
27. Ibid.
28. Ibid.
29. Ibid.
30. Ibid.
31. Ibid.
32. Ibid., July 16, 1958.
33. Ibid., Sept. 7, 1958.
34. Ibid.
35. Ibid., Sept. 24, 1958.
36. Poulson, *Memoirs,* pp. 206–7.
37. *City of Los Angeles v. the Superior Court of the County of Los Angeles.* 51 C. 2d
    423 (1959).
38. Ibid., p. 430.
39. Ibid.
40. Ibid., p. 432.
41. Ibid., p. 433.
42. Ibid., pp. 433–34.
43. Ibid., p. 434.
44. Ibid., p. 435.
45. Ibid., p. 436.
46. Ibid., p. 437.
47. Ibid.
48. Ibid., p. 438.
49. *Los Angeles Times,* Jan. 14, 1959.
50. Ibid.
51. Ibid.
52. Ibid.
53. Ibid., Jan. 15, 1959.

54. Ibid., Jan. 14, 1959.
55. Ibid., Jan. 15, 1959.
56. Arneburgh interview.
57. Information supplied by the Los Angeles Dodgers.
58. *Sporting News,* Feb. 11, 1959.
59. *Los Angeles Times,* Dec. 5, 1958.
60. Ibid., Apr. 9, 1959.
61. Ibid.
62. Ibid., Apr. 13, 1959.
63. Ibid., Apr. 9, 1959.
64. Kahn, *Boys of Summer,* pp. 323–26, and Tygiel, *Great Experiment.*
65. *Los Angeles Times,* Apr. 12, 1959. All subsequent accounts of Dodgers games for the 1959 season are taken from the *Los Angeles Times.*
66. Ibid., Apr. 16, 1959.
67. Ibid., May 9, 1959.
68. Ibid.
69. Poulson, *Memoirs,* p. 209.
70. *Los Angeles Times,* June 4, 1978.
71. Ibid., May 9, 1959.
72. Ibid., May 10, 1959.
73. Ibid.
74. Ibid., June 4, 1978.
75. Poulson, *Memoirs,* p. 211.
76. Ibid., p. 209.
77. Ibid., p. 212.
78. Poulson, *Memoirs,* pp. 211–12.
79. *Los Angeles Times,* May 14, 1959.
80. Ibid., May 15, 1959.
81. Author interview with Congressman Edward Roybal, Mar. 18, 1986.
82. Arneburgh interview.
83. Hahn interview.
84. *Los Angeles Times,* June 16, 1959.
85. Ibid., Sept. 2, 1959.
86. Ibid., Sept. 18, 1959.
87. Ibid., Oct. 9, 1959.
88. *New York Times,* Oct. 9, 1959.

Chapter 9

1. Author interview with Bowie Kuhn, Apr. 3, 1986.
2. Roger Noll, "The Economic Viability of Professional Baseball: Report to the Major League Players Association," July 1985.
3. Ibid., p. 31.
4. Ibid.
5. John Merwin, "The Most Valuable Executive in Either League," *Forbes,* April 12, 1982.
6. Roger Kahn, *A Season in the Sun* (New York: Harper and Row, 1977), p. 35.
7. Author interview with Peter O'Malley, Jan. 21, 1986, and Mar. 27, 1986.
8. Kuhn interview and O'Malley interview.

9. Kahn, *The Boys of Summer*, p. xii.

10. Golenbock, *Bums*, p. 446.

11. Ibid.

12. Ibid., p. 448.

13. Ibid., p. 432.

14. Ibid., pp. 438–39.

15. *New York Times*, Aug. 18, 1955.

16. *Brooklyn Eagle*, Dec. 11, 1953.

17. Kahn, *Boys of Summer*, pp. 334–36.

18. *Los Angeles Times*, Apr. 10, 1962.

19. Ibid., Mar. 30, 1966.

20. Ibid., Mar. 31, 1966.

21. Ibid.

22. Tommy Lasorda, *The Artful Dodger* (New York: Arbor House, 1985).

23. Scully interview.

24. See, e.g., Milton Friedman and Rose Friedman, *Free to Choose* (New York: Harcourt Brace Jovanovich, 1980).

25. See Nicholas Pileggi, "Was the Stadium Worth It?" *New York*, Apr. 19, 1976, pp. 36–42.

26. See Arthur T. Johnson, "Municipal Administration and the Sports Franchise Relocation Issue," *Public Administration Review*, Nov./Dec. 1983.

27. See "Professional Sports Antitrust Immunity." Hearings before the Committee on the Judiciary, U.S. Senate, 97th Cong., 2d sess.; and "Professional Sports Hearings," before the Subcommittee on Commerce, Transportation and Tourism of the Committee on Energy and Commerce, House of Representatives, 99th Cong., 1st sess.

28. *New York Times*, Dec. 18, 1985.

29. See Howard Cosell, *I Never Played the Game* (New York: William Morrow, 1985).

30. *New York Times*, May 9, 1982.

31. "The Impact of Professional Sports on the New York City Economy," Report of the Comptroller, the City of New York, Office of Policy Management, July 16, 1986.

32. *New York Times*, May 17, 1983.

33. See Ken Auletta, *The Streets Were Paved With Gold* (New York: Random House, 1979).

34. *New York Times*, Mar. 27, 1986.

35. See Peter Ueberroth et al., *Made in America: His Own Story* (New York: William Morrow, 1985).

36. *New York Times*, Dec. 3, 1985.

Index

Abrams, Cal, 16

Allen, Chester, 75

Allen, Mel, 33

Alston, Walter: in Brooklyn, 31-32, 58, 61-67; in Los Angeles, 153, 177, 185, 186-87, 196, 202; retirement of, 207-8

American Association, 5

American League, 6, 7, 41-42, 90-94

Angels, Los Angeles, 95, 107, 108, 128

Antitrust Committee (U.S. House of Representatives): and Dodgers' move to Los Angeles, 122-29; and exemption of baseball from antitrust laws, 120-29; and franchise relocation, 122-29; and Giants' move to San Francisco, 122-29; Giles's testimony before, 123; O'Malley's testimony before, 36, 56-57, 77, 95, 123-29; and Pacific Coast League, 93; and Parade Ground proposal, 126, 128; and pay television, 143; and reserve clause, 121-22; Stoneham's testimony before, 130

Arditto, James J., 120

Arechiga family, 178-81

Arneburgh, Roger: and Chavez Ravine agreement, 86, 87, 95, 99, 114; and Kirschbaum/Reuben lawsuits, 169, 170-71, 175, 181; and Proposition B, 146, 157-58

Athletics, franchise relocation of, 42-43

Atlantic-Flatbush proposal: Moses's objections to, 47-51, 54-57, 72-74, 75, 78-80, 97, 131-32; O'Malley's views about, 54-57. See also Brooklyn Sports Center

proposal; Clarke and Rapuano report; Madigan-Hyland report

Attendance: at Brooklyn Dodgers games, 20, 33, 40, 41, 43-44, 63, 68-69, 75-76, 77; at Los Angeles Dodgers games, 147, 148, 150-52, 178, 182, 190; and minor league proposals, 93-94

Baker, Dusty, 207

Baker, Earle, 160

Baltimore and franchise relocation, 6, 42, 44

Barber, Red, 10, 12, 33, 37, 41

Baseball: amateurism in, 4-5; and antitrust laws, 120-29, 214; as a business, 29, 53-54; cultural role of, 26; and drug/alcohol abuse, 209-10; as entertainment, 29, 121; and national/international issues, 9; organizational structure of, 8; scandals, 5-6, 8, 209-10; syndicated ownership in, 6, 23; and World War II, 89-90

Bavasi, E. J. (Buzzie), 138, 202-3

Baxes, Jim, 176

Beame, Abraham, 55

Bel Geddes, Norman, 37

Berkow, Ira, 15

Bessent, Don, 76

Bilko, Steve, 146-47, 148

Black, Joe, 25-26, 60, 62

Blacks, 18, 19, 21, 24-26, 31, 39

Blum, Robert, 75

Board of Estimate (New York City). See New York City administration

245